W9-BRM-481

A "Hands On" approach to teaching...

Geometry

Grades K-9

Ron Long   Linda Sue Brisby

Petti Pfau   Andy Heidemann

Scott Purdy   Natalie Hernandez

Sharon Rodgers   Jeanette Lenger

LEARNING ADVANTAGE   **FORT COLLINS, COLORADO**

Layout & Graphics: Scott Purdy
Primary Layout: Linda Sue Brisby and Jeanette Lenger
Cover Art: Petti Pfau and Jeff Barnes
Illustrations: Petti Pfau, Sharon Rodgers, Suzi Matthies

© 1989, 1993, 2007 Learning Advantage. All rights reserved.
Printed in the United States of America.

Limited Reproduction Permission: The authors and publishers hereby grant permission to the teacher who purchases this book, or the teacher for whom this book is purchased, to reproduce up to 100 copies of any part of this book for use with his or her students. Any other duplication is prohibited.

10 9 8 7 6 5 4

Product Number: 6040
ISBN-13: 978-0-927726-04-7
ISBN-10: 0-927726-04-1

LEARNING
ADVANTAGE

Learning Advantage
P.O. Box 368
Timnath, CO 80547

Contact us toll free **866-564-8251** or **info@learningadvantage.com**
to become a retail dealer or to locate a dealer near you!

Introduction

This book was compiled by a group of kindergarten through eighth grade teachers at Solvang Elementary School in Solvang, California. It is the fourth in a series of seven books that we are writing with the goal of filling an important niche in math education. Its purpose is to fill a void which we were experiencing at our school, and which we anticipated was being experienced at many other schools as well.

In 1985, the State of California released a document entitled, <u>Mathematics Framework For California Public Schools.</u> The document was revolutionary in that it sought to restructure the process of teaching math in the classroom. State commitment was so strong that all textbooks submitted for state adoption were initially rejected by the state textbook committee.

Classroom teachers were left in a state of confusion. Math strands, manipulatives, problem solving, cooperative learning, and calculators were "IN" (emphasized). Algorithms, memorization, pencil and paper math, and standardized tests were "OUT" (de-emphasized)! All of this occurred without the help of textbooks to bridge the transition.

At Solvang School we have been involved in a "hands-on" approach to math for a number of years. Still, we were caught in the situation of wondering how to apply the directions of the Framework.

To fill this void, we have created this set of activity books. It is not a textbook, but it is an invaluable supplement to your mathematics program.

All lessons described in this book are "activity based." We feel strongly that children learn best when thay have concrete experiences in learning mathematical concepts.

Our approach is to provide a TASK ANALYSIS of the skills children need to understand GEOMETRY, and to give a variety of activities which allow children to learn these skills. Activities are organized from basic to complex within each task analysis item, and each lesson has a list of necessary materials, recommended classroom organization, and a basic explanation of the lesson format. We have also included extensions in many lessons and have provided numerous experiences in teaching the metric system.

This book was written BY TEACHERS FOR TEACHERS, and we use these activities in our classrooms every day. All activities involve the use of easily obtainable and inexpensive objects as manipulatives. There is no need to spend large sums of money to teach math. We also feel that this approach enhances the "real world" applications of our lessons. We have left out the typical flow charts, color coding, and cross reference pages that often accompany multi-grade level texts. We have included only practical, teacher based information that you can read once and use.

What we have provided is over 150 organized, concise, activity oriented lessons for teaching GEOMETRY. We have included kindergarten through ninth grade lessons so any elementary teacher will have access to remedial and enrichment lessons in one convenient source. A second benefit of this format is that teachers can see the whole spectrum of the material that elementary age children should be learning.

About the Authors...

Linda Sue Brisby   M.A. – Indiana University; B.A. – Goshen College; Reading Specialist – University of Michigan; 19 years of teaching experience in primary grade levels.

Andy Heidemann   B.A. – Ohio Wesleyan University; 6 years of teaching experience in upper elementary and junior high.

Natalie Hernandez   B.A. – University of Illinois; 22 years of teaching experience in various middle grade levels.

Jeanette Lenger   B.S. – Cal Poly State University, San Luis Obispo; 23 years of teaching experience in primary grade levels.

Ron Long   M.A. – Cal Poly State University, San Luis Obispo; B.S. – Long Beach State University; 21 years of teaching experience at the junior high level.

Petti Pfau   M.A. – Cal Poly State University, San Luis Obispo; B.A. – San Francisco State University; 17 years of teaching in middle elementary grades and in special education.

Scott Purdy   M.A. – Western State College, Colorado; B.A. – UCLA: 17 years of teaching/administration from middle elementary through high school level.

Sharon Rodgers   B.S. – Southern Oregon College; 26 years of teaching experience at all elementary through junior high levels.

Table of Contents

Geometry
Task Analysis

Two dimensional	1. Compares, contrasts, and classifies circles, triangles, squares, and rectangles.	Primary
Two dimensional	2. Relates objects in the environment to geometric figures (squares, circles, triangles, and rectangles).	Primary
Two dimensional	3. Folds figures to show symmetry.	Primary
Three dimensional	4. Matches solid shapes (cube, cone, cylinder, spheré, rectangular prism) to objects in the room.	Primary
Two dimensional	5. Identifies simple congruent figures.	Primary
One dimensional	6. Recognizes angles (corners) in the room.	Primary
Relationships	7. Uses prepositions to describe relationship between two or more items.	Primary
Two dimensional	8. Identifies open and closed figures.	Primary
One dimensional	9. Identifies and names point and line.	Primary
One dimensional	10. Identifies and draws horizontal, vertical, and diagonal lines.	Primary
One dimensional	11. Identifies relationships and constructs representations of various one dimensional figures, their symbols, and their properties: point, line, line segment, ray.	Middle
One dimensional	12. Identifies relationships and constructs representations of various one dimensional elements and their symbols: intersection, bisection, parallel, and perpendicular.	Middle
One dimensional	13. Identifies relationships and constructs representations among one dimensional elements: coordinate/axis points on a coordinate graph or grid.	Middle
One dimensional	14. Identifies and labels angles according to three lettered points.	Middle
One dimensional	15. Identifies properties of angles: right, acute, and obtuse.	Middle
Two dimensional	16. Identifies properties of various triangles: base and height: scalene, isosceles, and equilateral.	Middle
Two dimensional	17. Identifies properties of various polygons: parallelograms, trapezoids, rhombuses, squares, and rectangles.	Middle
Two dimensional	18. Identifies and labels properties of a circle: diameter, chord, circumference, and radius.	Middle
Two dimensional	19. Constructs various two dimensional figures: polygons (parallelograms, rhombuses, trapezoids), circles, triangles (scalene, isosceles, equilateral).	Middle

Two dimensions	20. Constructs various two dimensional figures to show congruency, ratio, and proportion: polygons, circles, and triangles.	Middle
Symmetry	21. Determines lines of symmetry and investigates transformations in two dimensional figures.	Middle
Three dimensions	22. Identifies various three dimensional figures (prisms, cubes, spheres, cones, cylinders, polyhedra, etc.).	Middle
Three dimensions	23. Identifies the properties of various three dimensional figures: length, width, height, side/face, base, vertical, edge, volume, and surface area.	Middle
Multi-dimensions	24. Sorts and categorizes figures and their elements into the appropriate dimension (one, two, or three).	MIddle
One dimension	25. Identifies elements and constructs representations of one dimensional elements and their symbols: point, line, line segment, ray, arc, and angles.	Upper
One dimension	26. Identifies and constructs relationships among one dimensional elements: coordinate/axis points on a coordinate graph or grid.	Upper
Two dimensions	27. Identifies simple two dimensional elements and their symbols: angles (three letter notation), plane, and triangle.	Upper
Two dimensions	28. Identifies various complex two dimensional elements and their properties: circles, polygons, trapezoids, parallelograms, and rhombuses.	Upper
Two dimensions	29. Identifies relationships among simple two dimensional figures by analyzing vertex; acute, obtuse, right angles; complementary, supplementary and adjacent angles; diagonal, parallel, and perpendicular lines.	Upper
Two dimensions	30. Identifies relationships of complex two dimensional figures by analyzing vertex, acute, obtuse, and right angles; diameter and radius; congruency; interior angles.	Upper
Symmetry	31. Identifies and constructs various types of symmetry, transformations, and tesselations.	Upper
Ratio/Proportion	32. Identifies relationships among two dimensional figures to show ratio and proportion.	Upper
Two dimensions	33. Uses various methods (compass, protractor, and straight edge) to draw two dimensional figures.	Upper
Three dimensions	34. Identifies various three dimensional figures and their properties: sphere, cone, cylinder, polyhedra (cube, pyramid, prisms).	Upper
Three dimensions	35 Identifies relationships among various three dimensional elements: ratio and proportion, congruency, symmetry, base, edge, vertices, and side/face.	Upper
Three dimensions	36. Uses various methods to construct three dimensional shapes.	Upper
Multi-dimensions	37. Sorts, classifies, and constructs various figures into one, two, and three dimensions.	Upper

Fruity Lines
September
Grade Level: Primary

TASK ANALYSIS: 3 – Folds figures to show symmetry

MATERIALS: Citrus fruit: kiwi, orange, lime, lemon, grapefruit; picture examples of symmetry: "Vega–Nor" by Vasarely, "Pie Counter" by Thiebaud, "Three Musicians" by Picasso; knives, drawing paper, pencil, crayons

ORGANIZATION: K–3
Whole class activity for presentation
Children work in cooperative groups of 4–6 for task

PROCEDURE: – Show art prints and other examples of symmetry to the children.
– Point out symmetry.
– Have children find other examples in the prints and in the classroom.
– When several understand the concept, form cooperative groups.
– The cooperative groups will slice the fruit available and together they will discover the symmetry (pattern) of the interior of the slices.
– Groups will present their data to the whole class.
– Children record their favorite symmetrical slice of fruit by drawing.

– Extension: Extra fruit may be shared for snack time.
Explore a kaleidoscope.

lemon

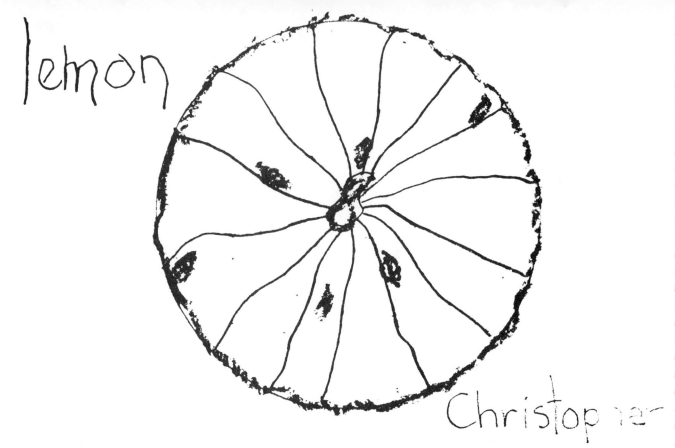

Christopher

grapefruit

Michele

2

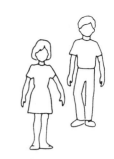

Beside – Between – Behind
September
Grade Level: Primary

TASK ANALYSIS: 7 – Use prepositions to describe relationship (location) between two or more items

MATERIALS: No extra materials needed

ORGANIZATION: K–3
Whole class activity

PROCEDURE:
- Have children work in groups of four.
- One child gives five directions such as: "Bobby, stand <u>behind</u> Tim;" "Jose, stand <u>beside</u> Tim;" "Bobby, put your hand <u>on</u> Jose's head;" "Tim, put your foot <u>between</u> Bobby and Jose."
- Change leaders.
- Do this procedure four times so that each child may be leader.

- Extension: On the playground, the teacher gives directives to the whole class in relationship to the playground equipment:
"Stand <u>beside</u> the swing."
"Sit <u>under</u> the slide."
"Sit <u>on</u> the see-saw."
"Stand <u>between</u> two swings."

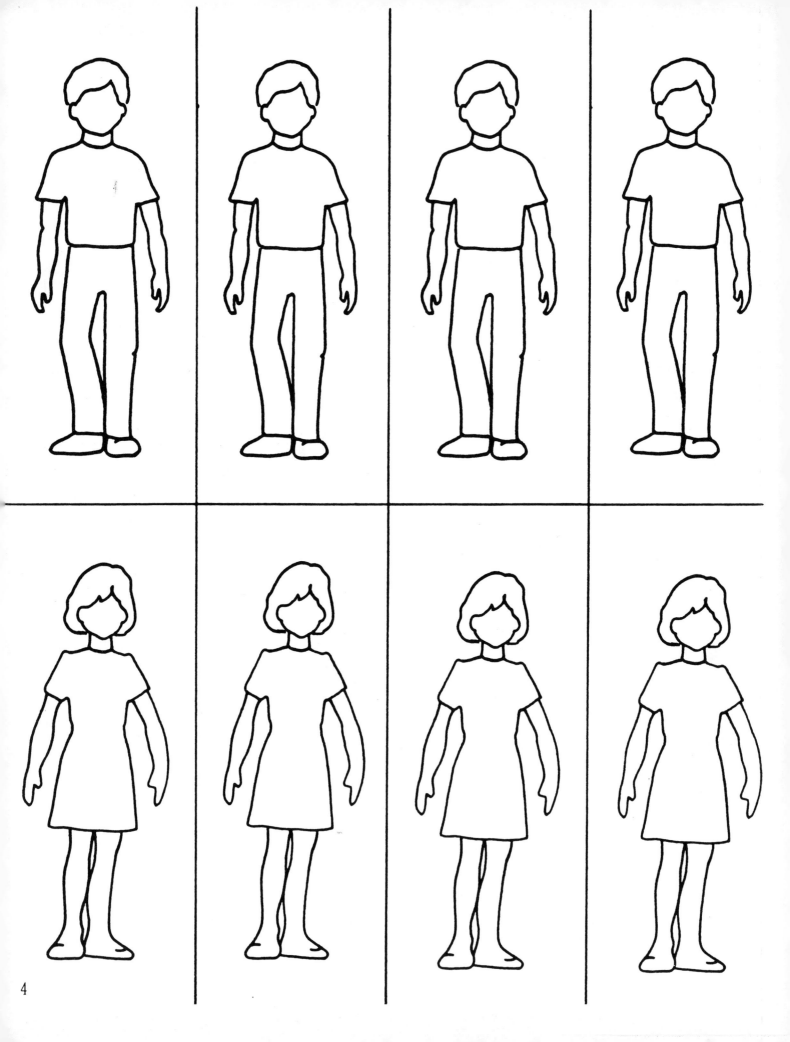

Point & Line No. 1
September
Grade Level: Primary

TASK ANALYSIS: 9 – Identifies and names point and line
10 – Identifies and draws horizontal, vertical, diagonal lines

MATERIALS: Art prints, large white paper or butcher paper, different colored wide markers, 12" x 18" (30 cm x 45 cm) white drawing paper (1 per child), pencils, crayons

ORGANIZATION: K–3
Whole class activity for presentation
Children do task individually

PROCEDURE:
- Using any art print available which shows line, gather the children close to the print. Ask the children to show any lines they may see in the print.
- Teacher may ask for straight lines, zigzag lines, curvy lines, etc.
- As one child traces a line on the print with his finger, ask others to "trace" in the air.
- On large drawing paper, the teacher makes two "dots" or "points" with a wide marker.
- Draw a "line" between the points. It may be straight, zigzag, or curvy, etc. Discuss.

- Draw two more "points" on the paper and draw a different line.
- Each time have the children "draw" the same line in the air.
- Continue this procedure until the children have seen examples of the following: straight, curvy, zigzag, horizontal, vertical, and diagonal lines.

– Give children a large piece of drawing paper, pencils and crayons.
– Have children explore by creating different lines by placing points on the paper and then drawing the lines between points.
– Children enjoy doing this lesson using a variety of medium. Watercolors produce very creative results.

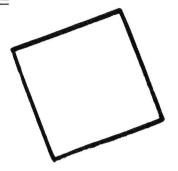

Attribute Block Congruency
September
Grade Level: Primary

TASK ANALYSIS: 5 – Identifies simple congruent figures

MATERIALS: Attribute blocks, 12" x 18" (30 cm x 45 cm) white drawing paper, pencils, crayons

ORGANIZATION: K–3
Whole class activity for presentation
Cooperative groups of 4–6 children

PROCEDURE: – Children free explore with attribute blocks in various positions to show congruency.
– Teacher models placement of triangles (or easiest attribute block) around border of placemat (white drawing paper).
– Children work in cooperative groups to create a border of congruent shapes on a placemat.
– Children may choose any attribute block to trace and color.
– Have children share the border they have created.

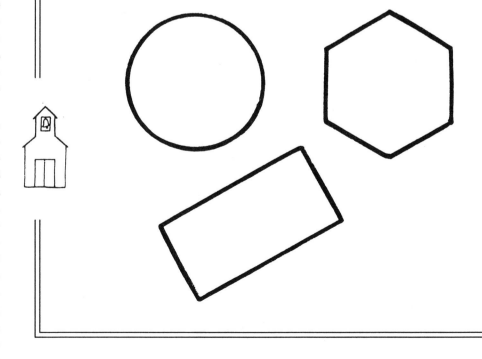

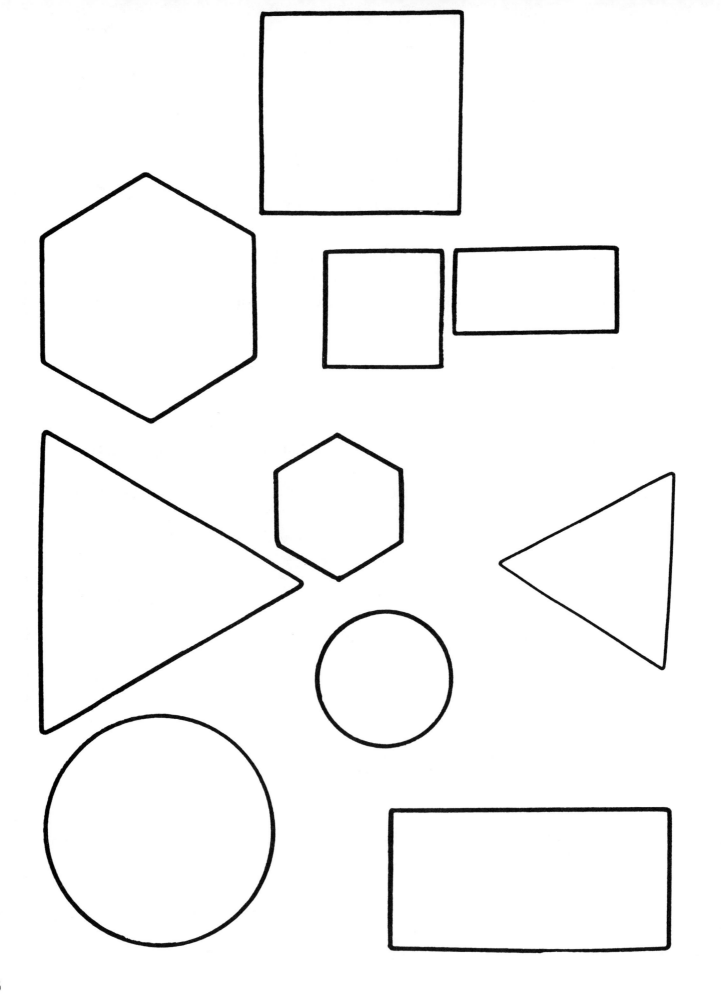

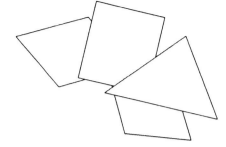

Pattern Block Attributes
October
Grade Level: Primary

TASK ANALYSIS: 5 – Identifies simple congruent figures

MATERIALS: Pattern blocks: hexagon, trapezoid, diamond, thin rhombus, 12″ x 18″ (30 cm x 45 cm) white drawing paper, pencils, crayons

ORGANIZATION: K–3
Whole class activity for presentation
Station activity with 2 children

PROCEDURE: – Children free explore with pattern blocks (if this is their first exposure to the blocks).
- Teacher models placement of hexagons around border of placemat (white drawing paper).
- Children work at station to create a border of congruent shapes on a placemat.
- Children may choose any pattern block.
- Have children share the border they have created.

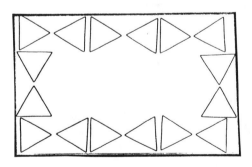

Pattern Blocks

A Solid Witch Hat
October
Grade Level: Primary

TASK ANALYSIS: 4 – Matches solid shapes (cube, cone, cylinder, sphere, rectangular prism) to objects
5 – Identifies simple congruent figures

MATERIALS: Tagboard shape patterns of triangle and rectangle, 12" x 18" (30 cm x 45 cm) construction paper placemats, pencils, crayons

ORGANIZATION: K–3
Whole class activity for presentation
2 children at a station to do task

PROCEDURE: – Teacher models tracing, cutting and constructing a witch hat using the triangle and rectangle patterns.
– Children create individual witch hats at station.
– Whole class creates a pattern of congruent shapes (witches hats) by taping them to blackboard or pinning to bulletin board.
– Teacher questions:
"What solid shapes do you see in a witch hat?"
"What other solid shapes could you use to make a witch hat?"

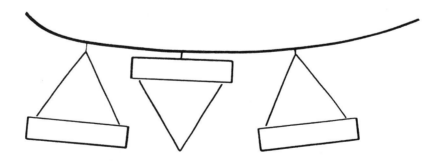

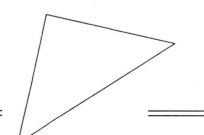

11

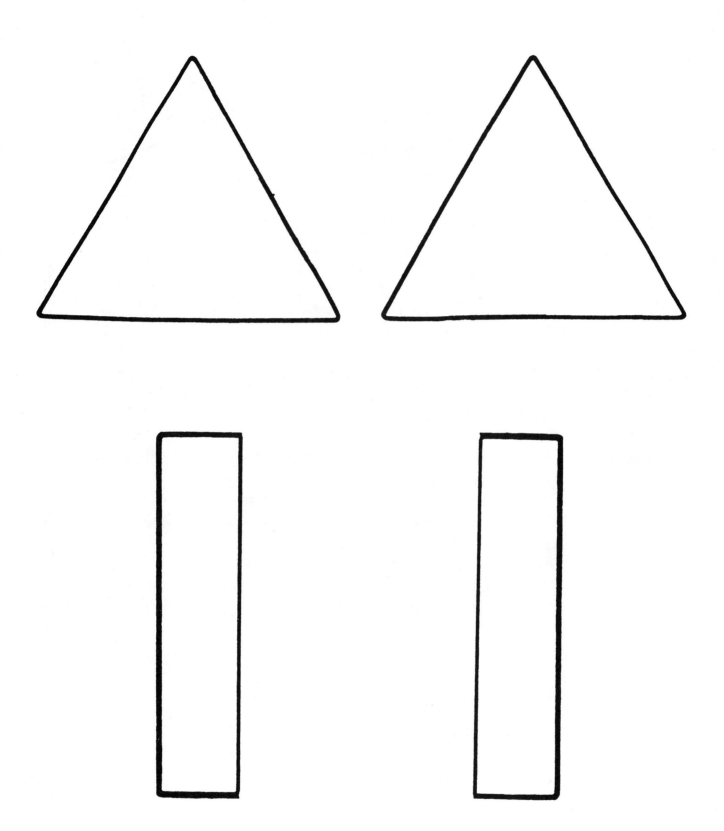

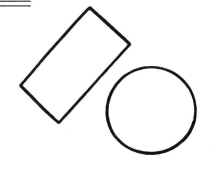

Attribute Block Attributes
October
Grade Level: Primary

TASK ANALYSIS: 1 – Compares, contrasts and classifies circles, triangles, squares, rectangles, trapezoids, hexagons, rhombus, and diamonds
6 – Recognizes angles (corners)

MATERIALS: Desk set attribute blocks per pair, set of large attribute blocks

ORGANIZATION: K–3
Whole class activity for presentation with children working in pairs for the task

PROCEDURE: – Children free explore with attribute blocks.
– As whole class (using large set of attribute blocks): have children discover point, line and angle of each attribute block.
– Children working in pairs (using desk set attribute blocks) do several sortings to discover point, line and angle.
– Discuss findings.
– Compare and contrast the data collected.

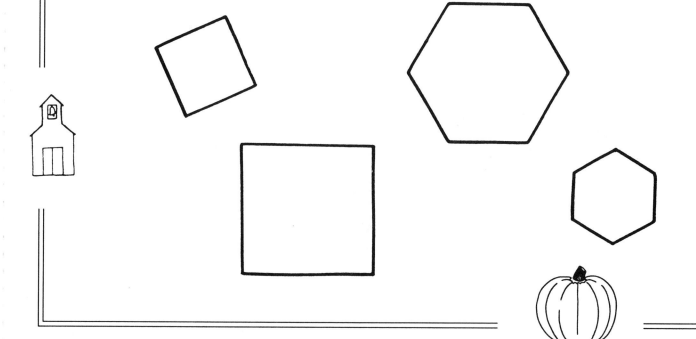

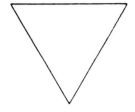

Attribute Block Data

Jack-O-Lantern Cut Outs
October
Grade Level: Primary

TASK ANALYSIS: 2 – Relates objects in the environment to geometric figures (squares, circles, triangles and rectangles)
3 – Folds figures to show symmetry
6 – Recognizes angles (corners)
9 – Identifies and names point and line

MATERIALS: Orange and black construction paper, paper, pencils, scissors, pictures of jack-o-lanterns, drawing paper, crayons

ORGANIZATION: K-3
Whole class activity for brainstorming
Individually for making jack-o-lanterns

PROCEDURE: – Show pictures of jack-o-lanterns discussing facial features.
– Have children draw a jack-o-lantern.
– Ask some children to share their creations describing the features they have given their jack-o-lantern.
– Teacher models the folding of the construction paper in half (either the long or the short way) and cut out a pumpkin shape and features of a jack-o-lantern.
– Discuss symmetry and shapes.
– Children then cut out their jack-o-lanterns.

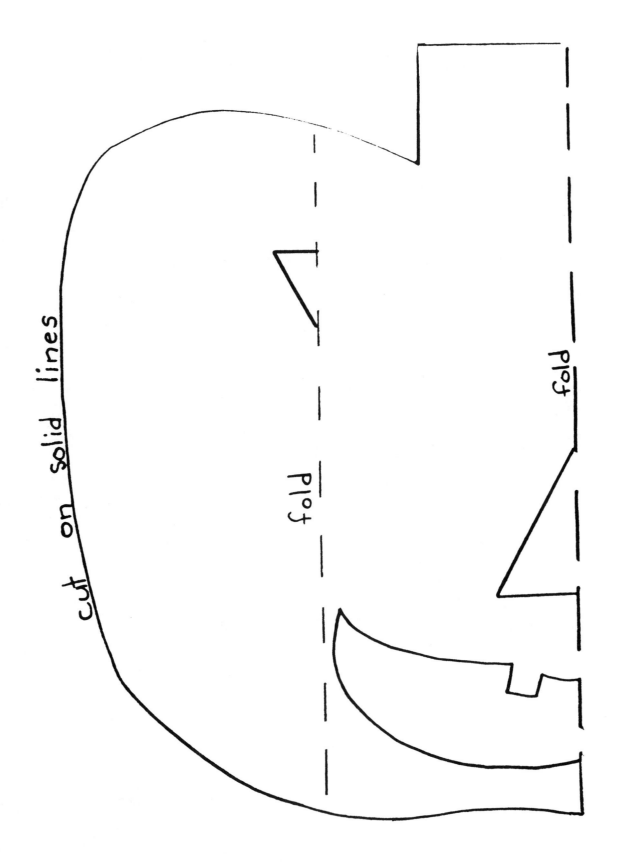

cut on solid lines

fold

fold

16

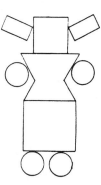

Rufus Robot
November
Grade Level: Primary

TASK ANALYSIS: 1 – Compares, contrasts and classifies circles, triangles, squares, rectangles, trapezoids, hexagons, rhombuses, and diamonds

MATERIALS: Attribute blocks, Rufus Robot page, sorting sheet with circle, square, triangle and rectangle, pencils, crayons.

ORGANIZATION: K–3
Whole class activity for presentation (use large attribute blocks). Children work in pairs for sorting and classifying (use desk attribute blocks).
Children work on Rufus Robot page individually.

PROCEDURE: – Sort and classify attribute blocks: 1) by shapes, 2) all 3-sided figures, all 4-sided figures, and all other figures.
– With circles, squares, triangles, and rectangles, create Rufus Robot.
– Record attribute blocks used, by coloring Rufus.

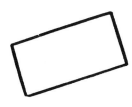 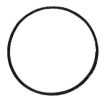 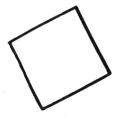

– For kindergarten, the presentation should be with the entire class using the large attribute blocks. For first grade and up, this activity may start with the desk attribute blocks.

– Extension: Compare Rufus to mammals.

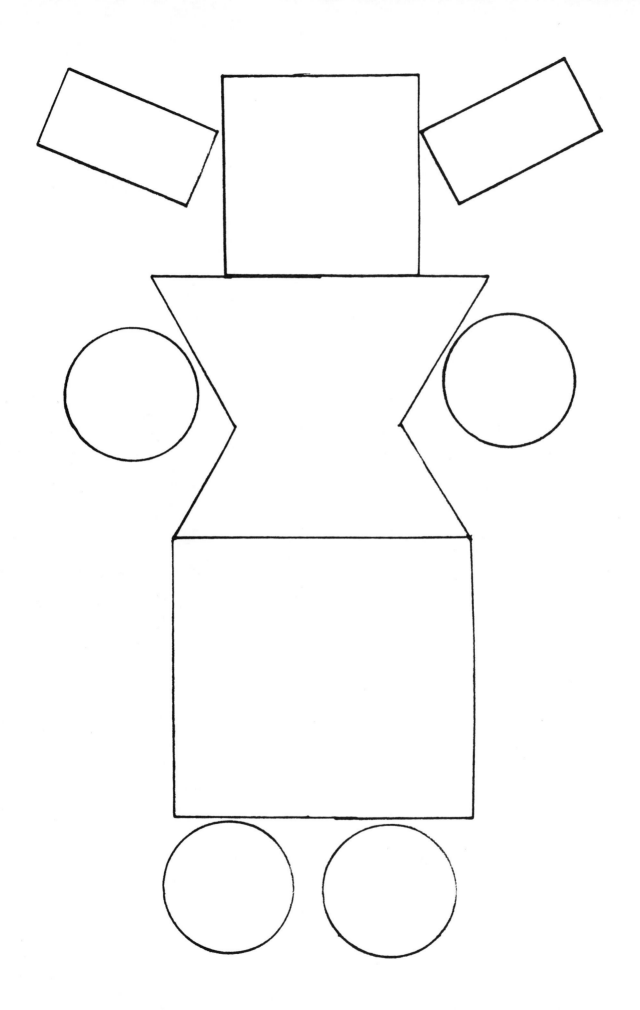

18

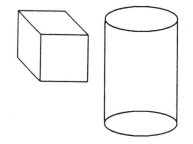

Solid, Solid, Solid
November
Grade Level: Primary

TASK ANALYSIS: 2 – Relates objects in the environment to geometric figures (squares, circles, triangles, and rectangles)
4 – Matches solid shapes (cube, cone, cylinder, sphere, rectangular prism) to objects

MATERIALS: Solid shapes cylinder (toilet paper roll), cone (ice cream cone), rectangular prism (kleenex box), sphere (ball), cube (block), paper pattern of each solid shape for children to construct, gathered items from home

ORGANIZATION: K–3
Whole class activity

PROCEDURE: – Construct solid shapes. Discuss.
– As a homework assignment, have the children bring an item from home to match the paper solid shapes.
– Have children sort and classify gathered items from home.
– Children use their own paper shapes as a guideline for classifying.
– Graph the various solid shaped items.
– Discuss collected data.

– Extension: Have the students go on a "solid shape" walk and take the constructed solid shapes to use as their guide.

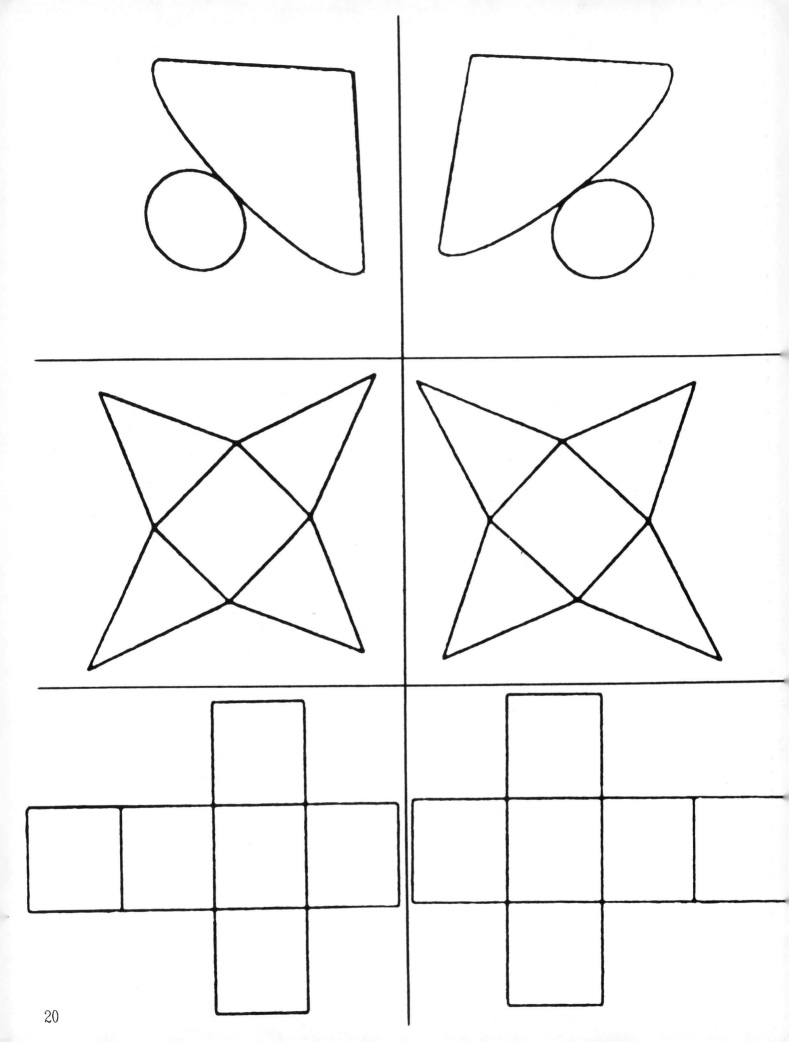

Falling Leaves
November
Grade Level: Primary

TASK ANALYSIS: 2 – Relates objects in the environment to geometric figures
(squares, circles, triangles, and rectangles)
3 – Folds figures to show symmetry
9 – Identifies and names point and line

MATERIALS: Fall–colored construction paper, pencils, scissors, crayons
(optional), autumn leaves, leaf patterns

ORGANIZATION: K–3
Whole class activity for presentation
Children do lesson individually

PROCEDURE: – Brainstorm with different leaves discussing shape, symmetry,
and vein patterns.
– Teacher models folding paper in half (either the long or short
way) and cutting a leaf.
– Use crayons or pencil to put in vein pattern discussing point and
line.

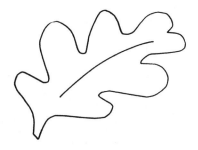

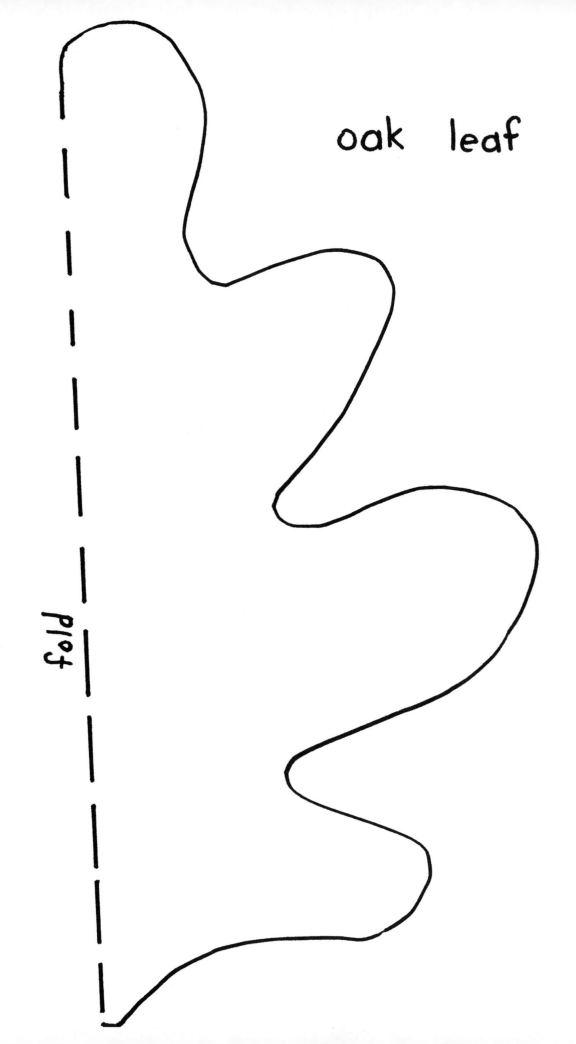

oak leaf

fold

22

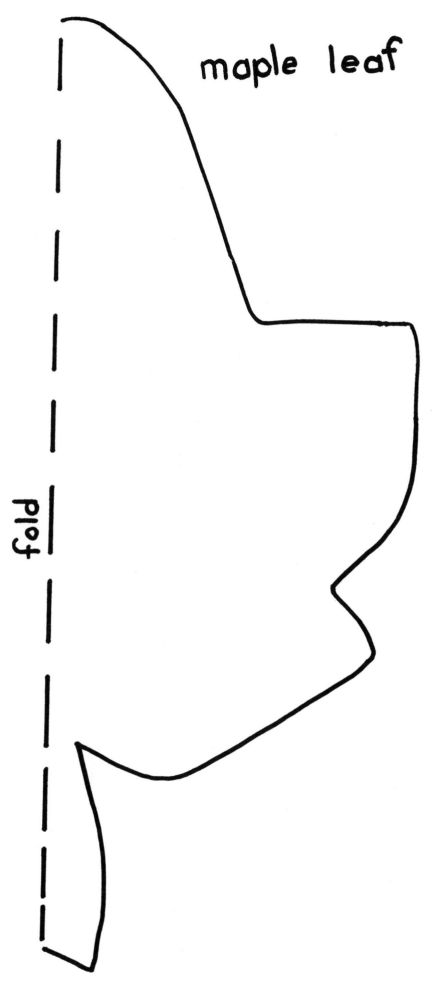

maple leaf

fold

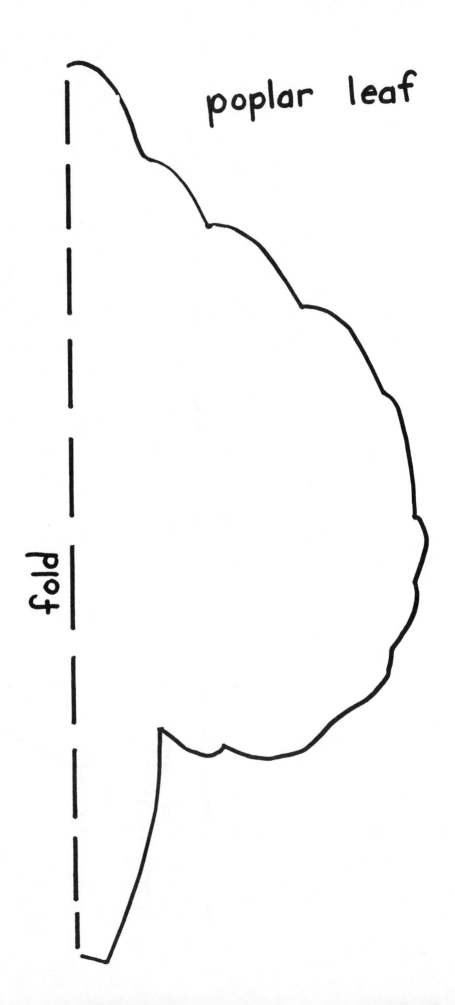

poplar leaf

fold

24

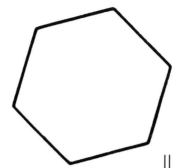

Pattern Block, What's Your Angle?
November
Grade Level: Primary

TASK ANALYSIS: 6 – Recognizes angles (corners)
9 – Identifies and names point and line

MATERIALS: Pattern blocks, sorting mats, big shapes of tagboard representing each pattern block

ORGANIZATION: K–3
Whole class presentation with children forming pairs to do task

PROCEDURE: – As whole class, discuss each pattern block shape discovering points, lines and angles.
– Children working in pairs, sort pattern blocks and discuss point, line, angle.
– Children need to sort pattern blocks several times to discover the many common attributes.

– Extension: Geometric Hop-Scotch: With the large tagboard shapes or chalk shapes play a hop-scotch game –– "Jump into the shape with 4 angles," etc.

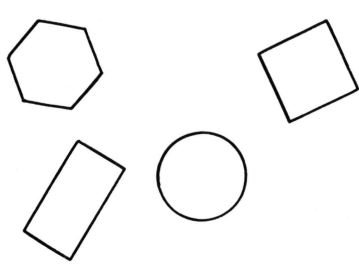

Pattern Blocks

Shape It: Open/Closed
November
Grade Level: Primary

TASK ANALYSIS: 8 – Identifies open and closed figures
9 – Identifies and names point and line
10 – Identifies horizontal, vertical, diagonal lines and draws

MATERIALS: 12" (30 cm) pieces of fat yarn of different colors (not attached), 1 per child

ORGANIZATION: K–3
Whole class activity

PROCEDURE: – Brainstorm for open and closed shapes: circle/arc, music triangle, curvy line, zigzag line, milk carton, aquarium, etc.
– Have children create open and closed shapes with their yarn.
– Children share their discoveries with classmates by identifying points, lines, kinds of line, and if an open or closed shape.
– Place the children into cooperative groups.
– Give them a longer piece of yarn.
– Have the children create a closed figure with three lines.
– Have the children create an open figure with curvy lines.
– Continue giving this type of direction.

diagonal curvy line.

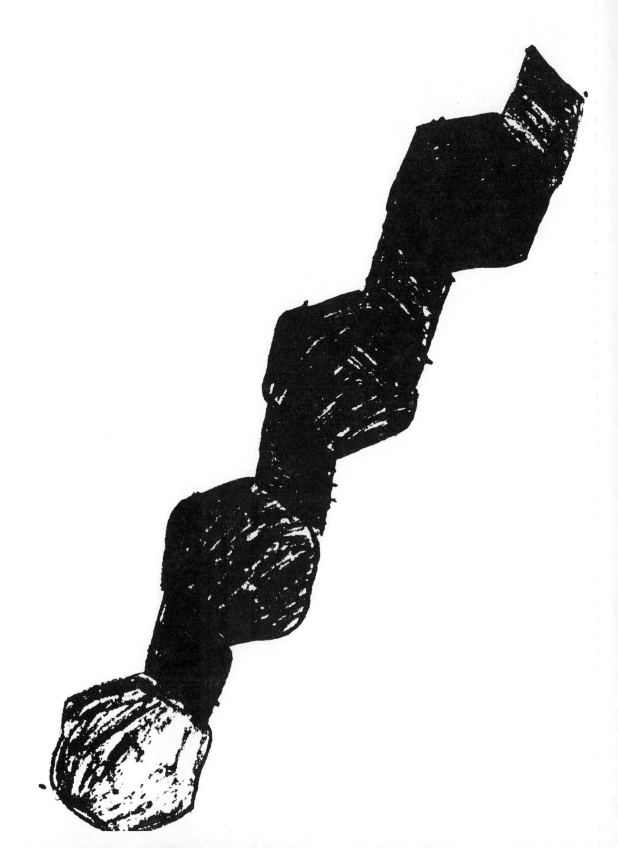

Point & Line No. 2
December
Grade Level: Primary

TASK ANALYSIS: 1 – Compares, contrasts and classifies circles, triangles, squares, rectangles, trapezoids, hexagons, rhombuses, and diamonds
9 – Identifies and names point and line
10 – Identifies and draws horizontal, vertical, diagonal lines

MATERIALS: Attribute and/or pattern blocks, 10" x 10" (25 cm x 25 cm) white drawing paper, pencils, crayons or fine point felt tip markers

ORGANIZATION: K–3
Whole class activity for presentation
Children do task individually
We strongly suggest the children experience Point & Line No. 1 before using this lesson.

PROCEDURE: – Give the children a 10" x 10" piece of drawing paper, attribute/pattern blocks and pencils.
– Child places two "dots" or "points" on the drawing paper.
– Next child places blocks in a line between the two dots.
– Child traces around each block in the line and colors in the shapes to form a color pattern.

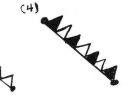

(1)   (2)   (3)   (4)

– Child names and labels line, (this could be straight–diagonal, zigzag–diagonal, etc.)
– Children like to create more than one "type" of line.

Draw five shapes.

Use the shapes you just drew to create something.

Draw a curvy diagonal line. Color it with a primary color.

Draw a horizontal zigzag. Color it with a secondary color.

Draw a straight vertical line. Color it your favorite color.

Gingerbread People
December
Grade Level: Primary

TASK ANALYSIS: 2 – Relates objects in the environment to geometric figures (squares, circles, triangles, and rectangles)
3 – Folds figures to show symmetry

MATERIALS: Large brown construction paper, example of gingerbread paper doll, scissors, pencils, crayons

ORGANIZATION: K–3
Whole class activity with teacher "walking" the children through the lesson

PROCEDURE: – Teacher models the following steps:
"Fold paper in half."
"Fold paper in half again in the same direction."

– Show how to draw gingerbread man on the paper.
– Cut on the lines.
– Open up and decorate, making "look–alike" gingerbread people.

Geoboard Geometry
December
Grade Level: Primary

TASK ANALYSIS: 2 – Relates objects in the environment to geometric figures (squares, circles, triangles, and rectangles)
6 – Recognizes angles (corners)
9 – Identifies and names point and line
10 – Identifies and draws horizontal, vertical, diagonal lines

MATERIALS: Geoboards, geobands

ORGANIZATION: K–3
Whole class activity with children working in pairs

PROCEDURE: – Show children different geometric shapes.
– Free explore making geometric shapes with geobands on geoboards.
– Each pair will tell the shapes they created, the number of angles in each shape, what they see in the classroom with the same shape and same number of angles, and the horizontal, vertical and diagonal lines.
– Children identify point and line in the shapes.
– Discuss the likenesses and differences of each pair's shapes.

– Extension: Create your classroom on the geoboard and record on geoboard paper.

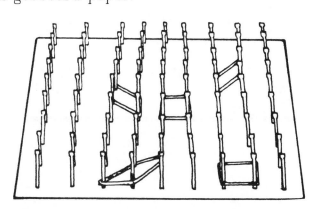

Evergreen Trees
December
Grade Level: Primary

TASK ANALYSIS: 4 – Matches solid shapes (cube, cone, cylinder, sphere, rectangular prism) to objects
5 – Identifies simple congruent figures

MATERIALS: Tree pattern, green construction paper, 12" x 18" (30 cm x 45 cm) white drawing paper (for placemat), assortment of evergreen tree pictures, crayons, pencils, scissors, glue

ORGANIZATION: K–3
Whole class activity for presentation
Station activity

PROCEDURE: – Teacher questions: "What solid shapes do you see in an evergreen tree?" (cylinder, cone)
– Teacher models placement of tree pattern on placemat.
– Children free explore with tree patterns.
– Children trace and cut out trees.
– Children create a border of congruent trees on their placemats.

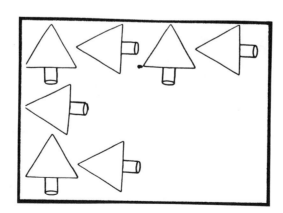

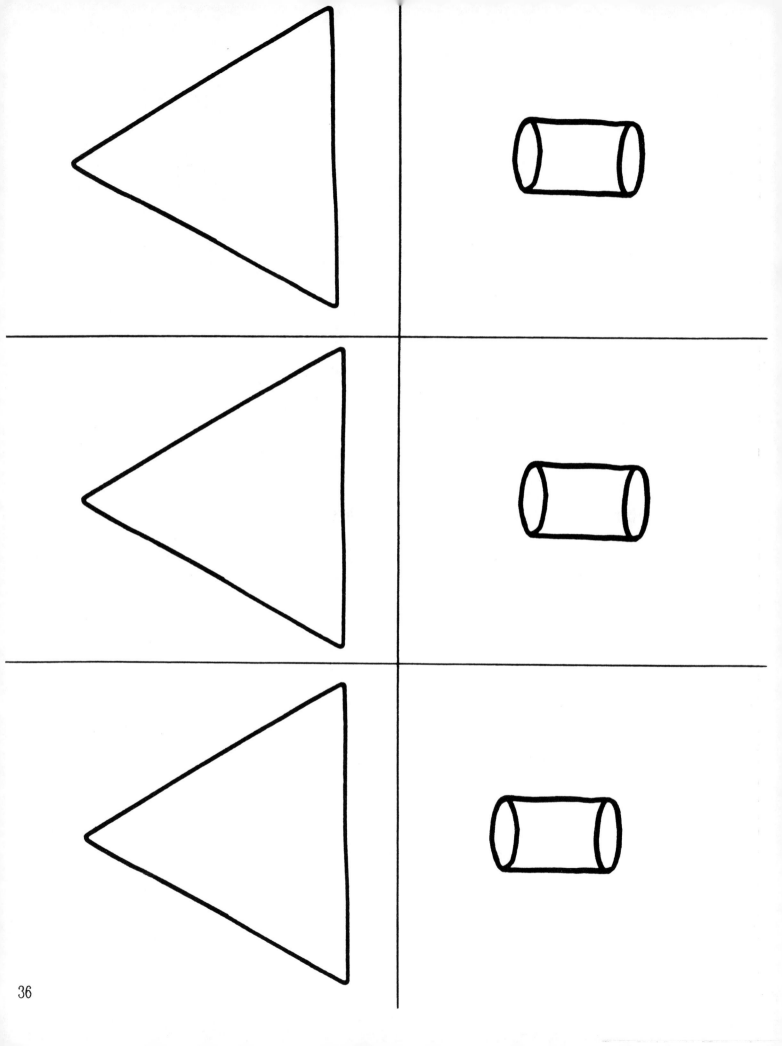

36

Geometry Toys
January
Grade Level: Primary

TASK ANALYSIS: 2 – Relates objects in the environment to geometric figures (square, circles, triangles, and rectangles)
4 – Matches solid shapes (cube, cone, cylinder, sphere, rectangular prism) to objects

MATERIALS: Toys brought from home, one floor graph, tent headings for first activity: square, rectangle, triangle, circle; Tent headings for second activity: cube, cone, cylinder, sphere, rectangular prism

ORGANIZATION: K–3
Cooperative groups of 4 – 6 children

PROCEDURE: – Use first set of headings.
– Children introduce and describe their toys to the class.
– Children place toys where they think they belong.
– Interpret data.
– Have one cooperative group arrange the toys according to their understanding.
– Report to the whole class.
– Follow same procedure for all cooperative groups.

– Set up the second set of headings (solid shapes).
– Children re-introduce toys and continue above procedure.

– Extension: Make butcher paper graph and have children record data.

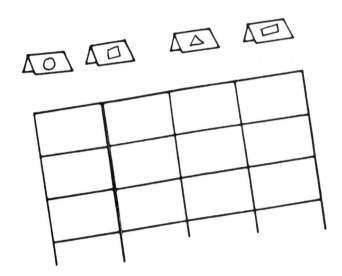

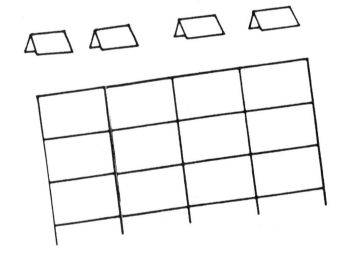

Trees, Trees, Trees
January
Grade Level: Primary

TASK ANALYSIS: 2 – Relates objects in the environment to geometric figures (squares, circles, triangles, and rectangles)
3 – Folds figures to show symmetry
10 – Identifies and draws horizontal, vertical, diagonal lines

MATERIALS: 9" x 12" (22.5 cm x 30 cm) green construction paper, 2 pieces per child, tree pattern to trace, scissors, crayons, stapler, pencils, tape

ORGANIZATION: K–3
Whole class presentation with teacher modeling
Children do activity individually

PROCEDURE: – Teacher models folding BOTH pieces of green construction paper in half.
– Trace the tree pattern and cut out.
– Staple folds together to cause a 3D effect.
– Children decorate tree symmetrically.
– Tape to middle of childrens' tables to make an evergreen tree lane.

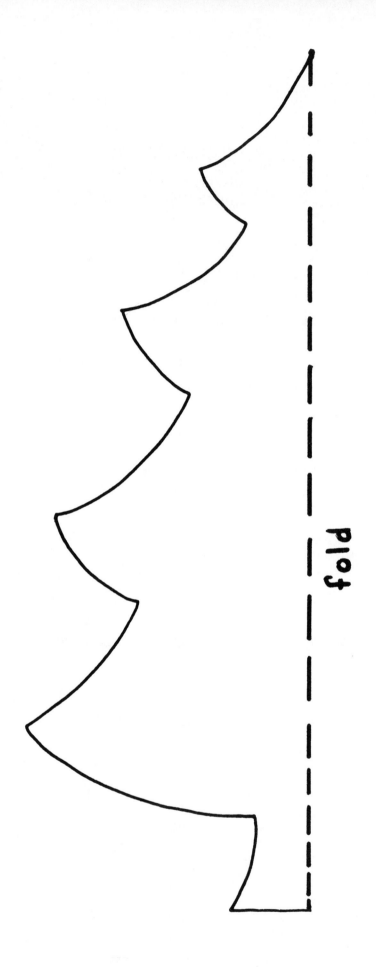

fold

40

Going to a Fire
January
Grade Level: Primary

TASK ANALYSIS: 7 – Use prepositions to describe relationship (location) between two or more items
9 – Identifies and names point and line
10 – Identifies and draws horizontal, vertical, diagonal lines

MATERIALS: Drawing paper 12" x 18" (30 cm x 45 cm), crayons

ORGANIZATION: K–3
Whole class activity, 15–20 minutes

PROCEDURE: – Tell children they're going to draw a fire truck.
– Have children fold their paper in half the long way (hot dog way!) and open it up.
– Teacher models while giving directions:
– Trace center fold line with your finger.
– Put your fingers on both end points.
– Now, put your fingers on two other points of the center fold line.
– Draw (with crayons) a line between these two points.
– Using two vertical lines and one horizontal line, make the engine.
– Teacher continues giving geometric directives using prepositions (i.e. – "Draw a square next to the engine.").

– Suggestion: Ed Emberly drawing books.

William

Ernesto

42

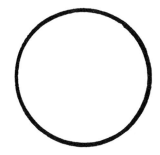

What's Your Angle?
January
Grade Level: Primary

TASK ANALYSIS: 6 – Recognizes angles (corners)
8 – Identifies open and closed figures

MATERIALS: Large pieces of butcher paper, one with no angles, one with 1 angle, one with 2 angles, one with 3 angles, one with 4 angles, and one with more than 4 angles, 4" x 4" (10 cm x 10 cm) paper 1 per child, crayons, glue

ORGANIZATION: K–3
Whole class activity

PROCEDURE: – Have butcher paper shapes hanging on the wall.
– Explain to the children that they are going outside to look for angles.
– Brainstorm for ideas of angles.
– Take an "angle walk."
– After the angle walk, children draw on 4" x 4" paper an example of angle(s) discovered on angle walk.
– Children tell what their picture is and attach it to the matching large shape (same number of angles) on the wall.
– Children tell if it is an open or closed figure.
– Children explain why.

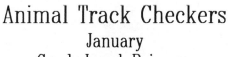

Animal Track Checkers
January
Grade Level: Primary

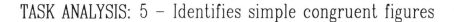

TASK ANALYSIS: 5 – Identifies simple congruent figures

MATERIALS: 12" x 18" (30 cm x 45 cm) white drawing paper, xeroxed copies of animal tracks, scissors, glue

ORGANIZATION: K–3
Whole class activity for presentation
Station activity

PROCEDURE:
- Brainstorm for different animal tracks.
- Read The Snowy Day by Ezra Jack Keats.
- Children use xeroxed sheet of animal tracks.
- Cut out needed tracks to make checkerboard pattern on the white drawing paper.
- Children share and discuss their congruent patterns.
- Laminate and use as placemat or gameboard.

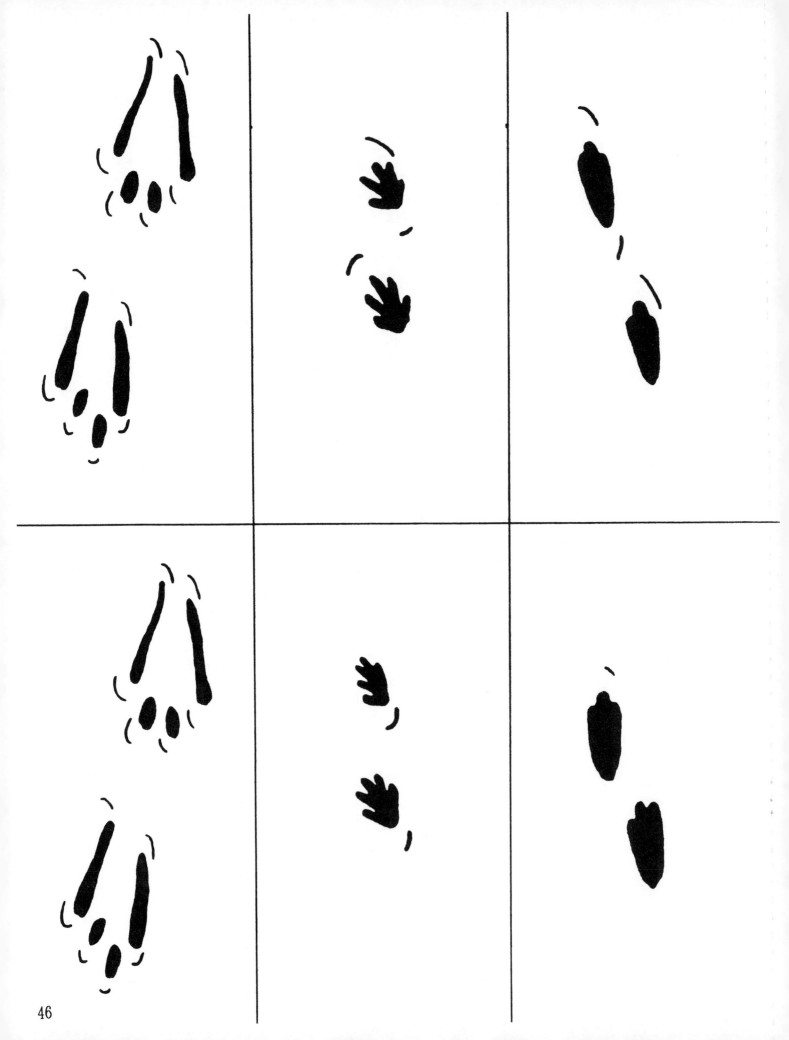

46

Lincoln ^{Lincoln} Washington ^{Washington}

February
Grade Level: Primary

TASK ANALYSIS: 5 – Identifies simple congruent figures

MATERIALS: Lincoln and Washington silhouette stamps, ink pads, 12" x 18" (30 cm x 45 cm) white drawing paper

ORGANIZATION: K–3
Station activity

PROCEDURE:
- Teacher shows some sample borders of congruent patterns.
- Children free explore with stamps to create congruent patterns.
- Children stamp Lincoln up and Lincoln down, Washington up and Washington down around the entire border.

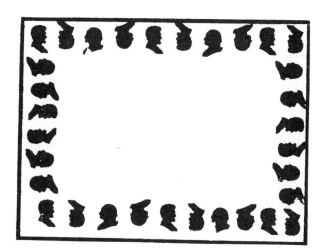

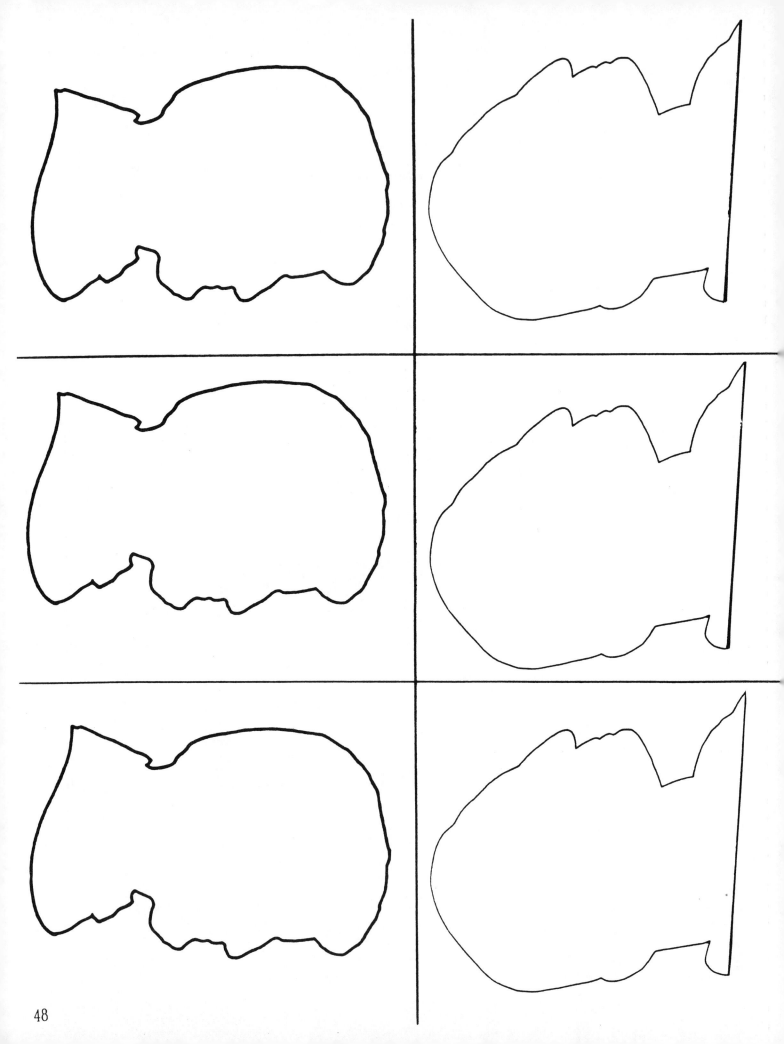

48

Three Corners Makes a Hat
February
Grade Level: Primary

TASK ANALYSIS: 2 – Relates objects in the environment to geometric figures (square, circles, triangles, and rectangles)
5 – Identifies simple congruent figures
6 – Recognizes angles (corners)
8 – Identifies open and closed figures

MATERIALS: Red, white and blue construction paper, hat pattern, bow pattern, 4" x 6" (10 cm x 15 cm) white paper (for hair), pencil, scissors, stapler

ORGANIZATION: K – Whole class activity
1-3 – Whole class presentation, then a station activity

PROCEDURE: – Pattern can already be traced on colored construction paper, or children may do the tracing.
– Trace the pattern on rectangular pieces of construction paper (red, white and blue) and cut out.
– Trace the bow pattern on a smaller rectangular piece of blue construction paper, and cut out.
– Fringe the 4" x 6" white paper for hair and curl on a pencil.

– Finish making the hats by stapling the hat pieces together. Add the hair and bow to bottom of blue piece of the hat (back of hat).
– Discuss the shapes used for making the hat (rectangles for the sides, two triangles for the bow, the finished hat is a triangle, the hat has three angles, it started as an open figure and when it was finished, it was a closed figure).

– Extensions: Second and third grades may read books, create stories and poems about George Washington.

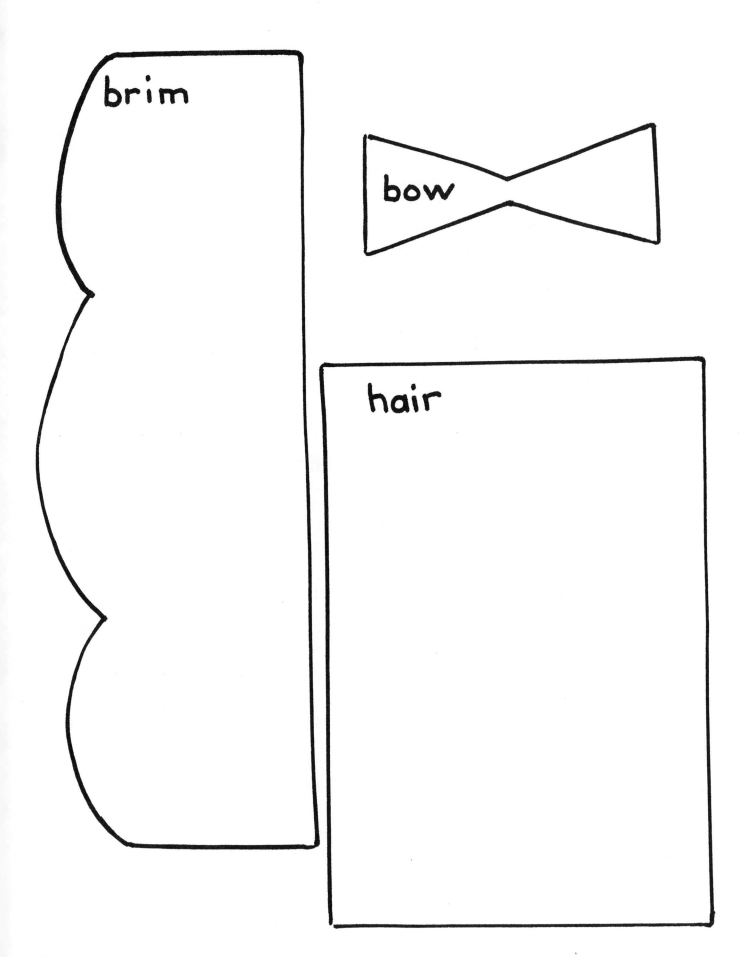

brim

bow

hair

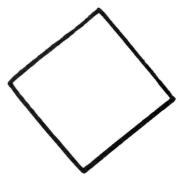

On–Off–Under

TASK ANALYSIS: 7 – Use prepositions to describe relationship (location) between two or more items

MATERIALS: Pattern blocks and/or attribute blocks, working mats

ORGANIZATION: K–3
Whole class activity with children working in pairs

PROCEDURE:
- Brainstorm for prepositions.
- Place prepositions in pocket chart.
- Children work in pairs at their desks.
- Give each pair (amount) pattern/attribute blocks.
- Children will take turns giving a prepositional direction:
  1st child – "Put your triangle in the middle of the mat."
  2nd child – "Put your circle beside the triangle."
  1st child – "Put your small square on top of the circle."

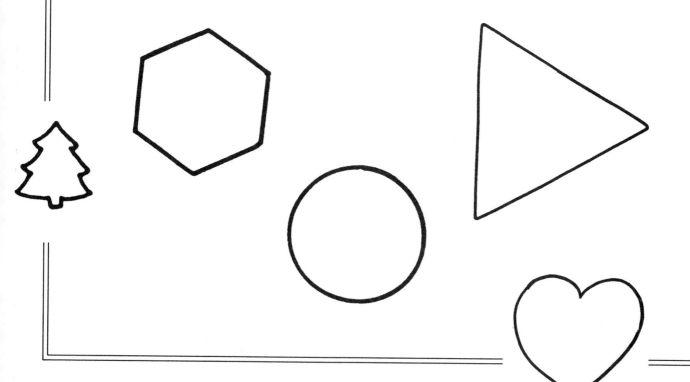

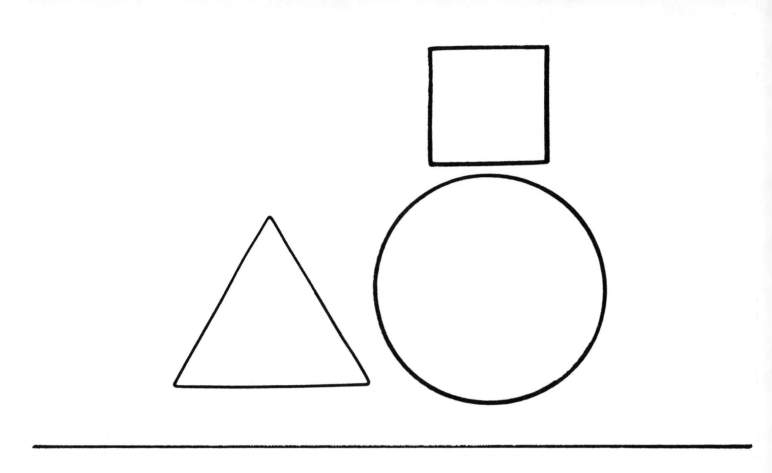

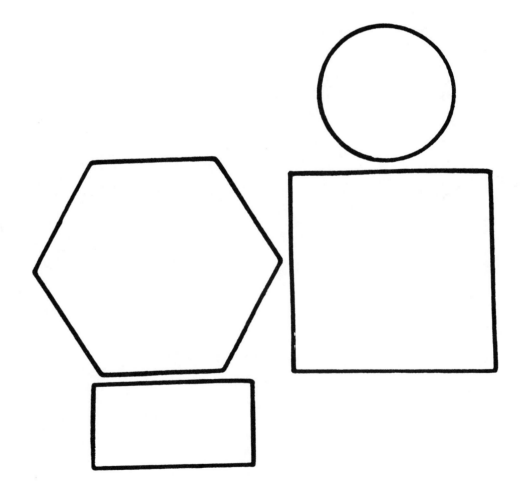

52

Hearts–Hearts
February
Grade Level: Primary

TASK ANALYSIS: 3 – Folds figures to show symmetry

MATERIALS: Colored construction paper (any size), scissors, pencils

ORGANIZATION: K–3
Whole class for presentation
Children do lesson individually

PROCEDURE: – Teacher models the following procedures:
- Fold the construction paper in half.
- Draw one half of a heart and cut out the heart.
- Draw lines where other cuts will be made. Cut out darkened portions of heart.

Children follow same procedure.
Hearts may be hung around the room or mounted on contrasting colored paper.

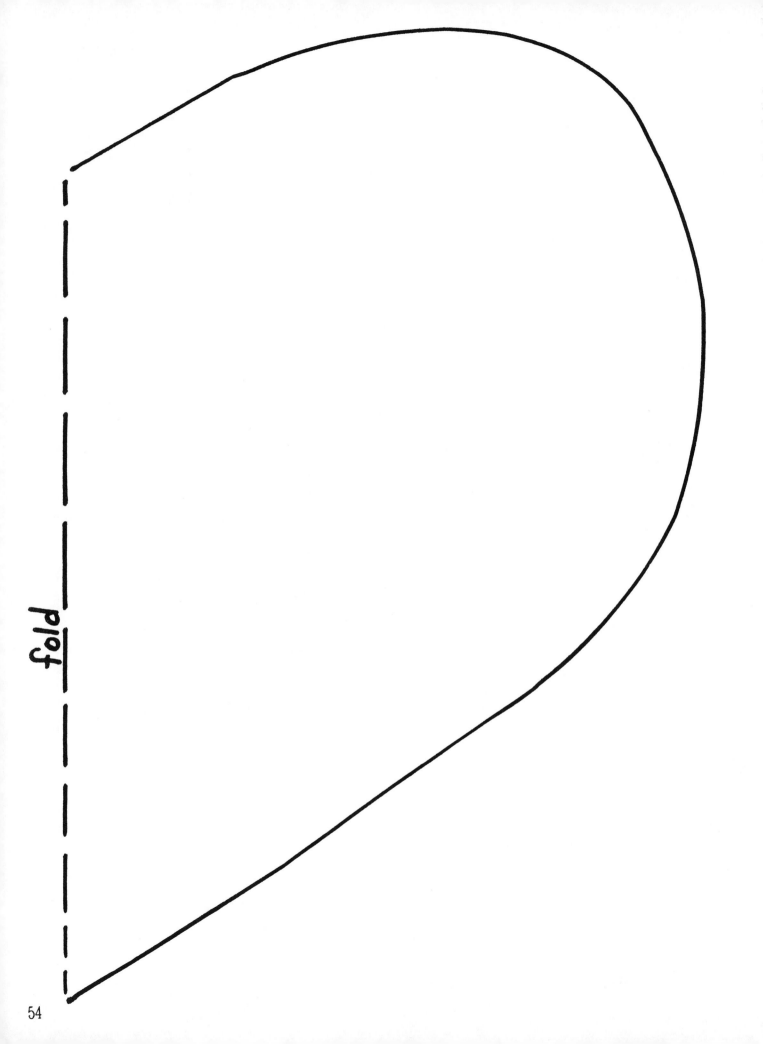

fold

54

Is Our Room in Solid Shape?
February
Grade Level: Primary

TASK ANALYSIS: 4 – Matches solid shapes (cube, cone, cylinder, sphere, rectangular prism) to objects
6 – Recognizes angles (corners)
9 – Identifies and names point and line
10 – Identifies and draws horizontal, vertical, diagonal lines

MATERIALS: Examples of the solid shapes – rectangular prism, cube, cone, cylinder, sphere

ORGANIZATION: K–3
Whole class activity, children working in cooperative groups of 4–6

PROCEDURE: – Present solid shapes to children.
– Let them explore the shapes finding points, lines (horizontal, vertical, diagonal), and angles.
– Brainstorm ways to explore the classroom to find solid shapes.
– From brainstormed plan, have children take a solid shape walk in the classroom to discover solid shapes.
– As a whole class, discuss findings.

– Extension: Graph findings and interpret data.

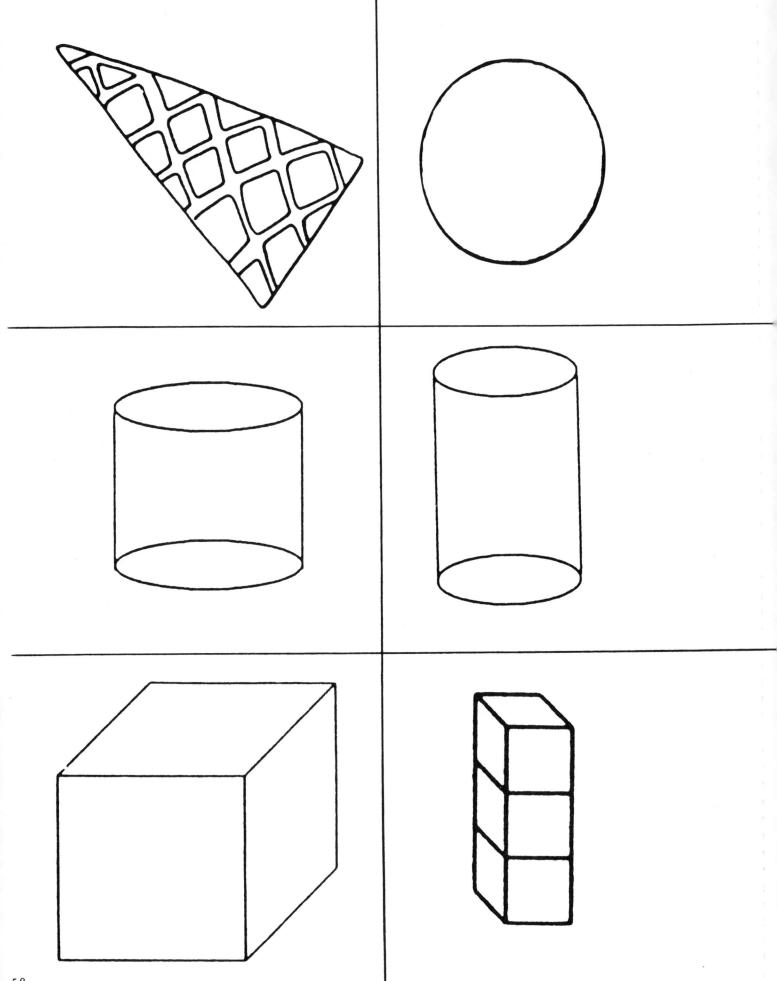

Seeing Double
March
Grade Level: Primary

TASK ANALYSIS: 2 – Relates objects in the environment to geometric figures
(squares, circles, triangles, and rectangles)
7 – Use prepositions to describe relationship (location) between
two or more items
8 – Identifies open and closed figures
9 – Identifies and names point and line
10 – Identifies and draws horizontal, vertical, diagonal lines

MATERIALS: Graph paper with designs created from geometrical shapes,
pencils, crayons

ORGANIZATION: K–3
Whole group activity for presentation
Children do lesson individually

PROCEDURE: – Brainstorm (words from above).
– Using the various lines and shapes of the sample graph paper
designs, children try to duplicate these designs on their own
graph paper.
– Children need to try several of the different designs.
– Discuss relationships of shapes in their pictures (i.e. – "The
circle is above the square.").
– Students should identify open and closed figures.

– Extension: Children may create original designs and figures on
graph paper, using their mastery of point, line and geometric
shapes.

57

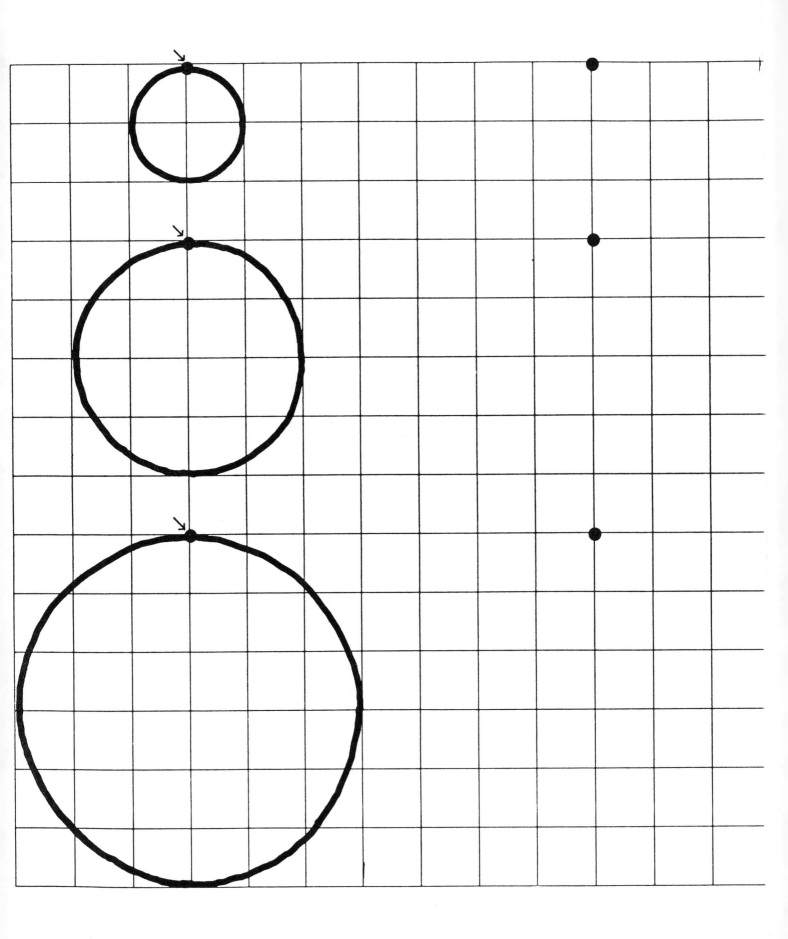

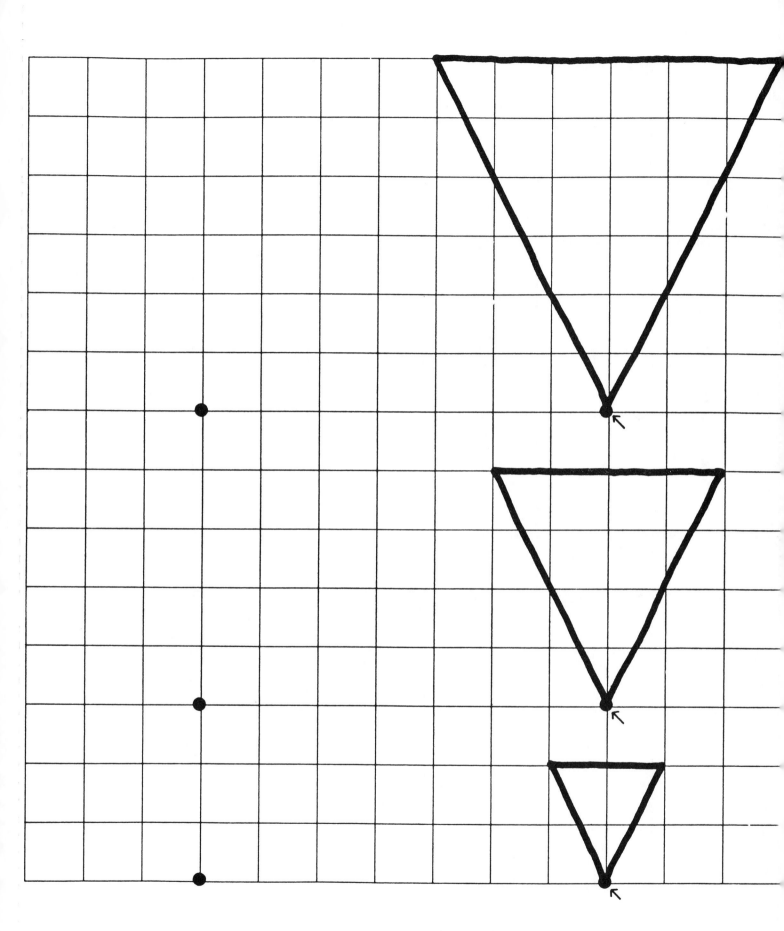

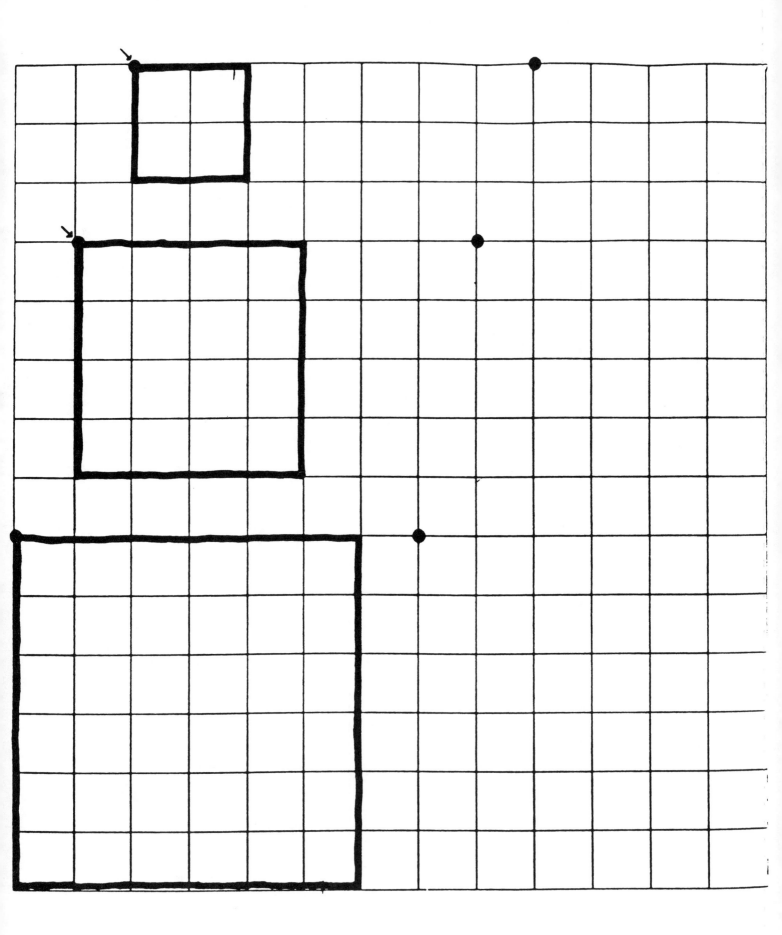

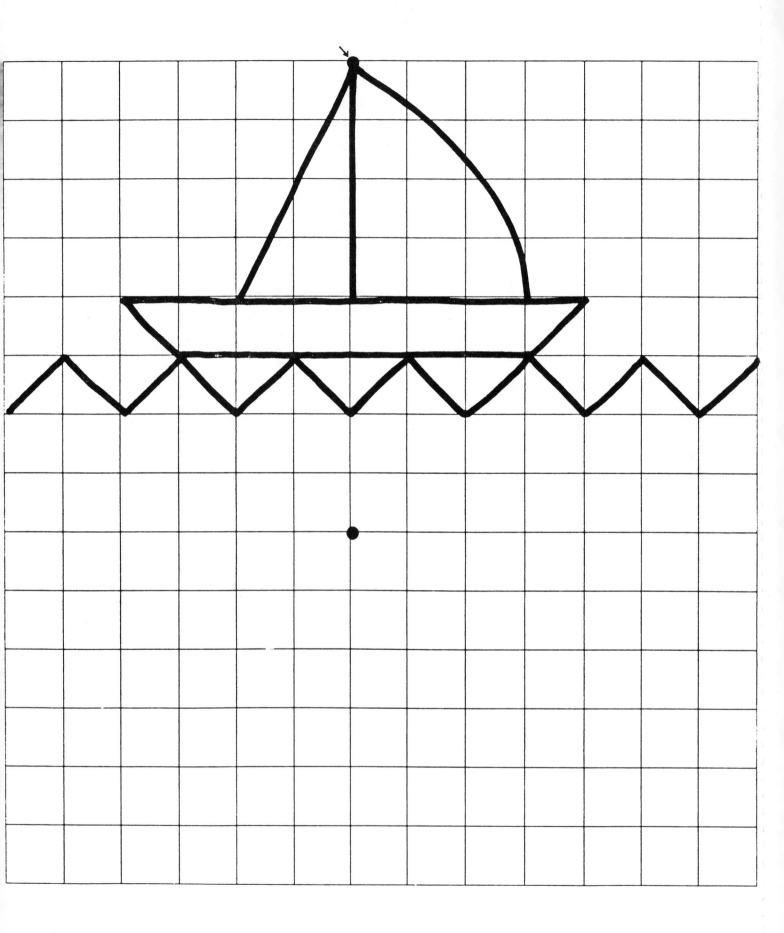

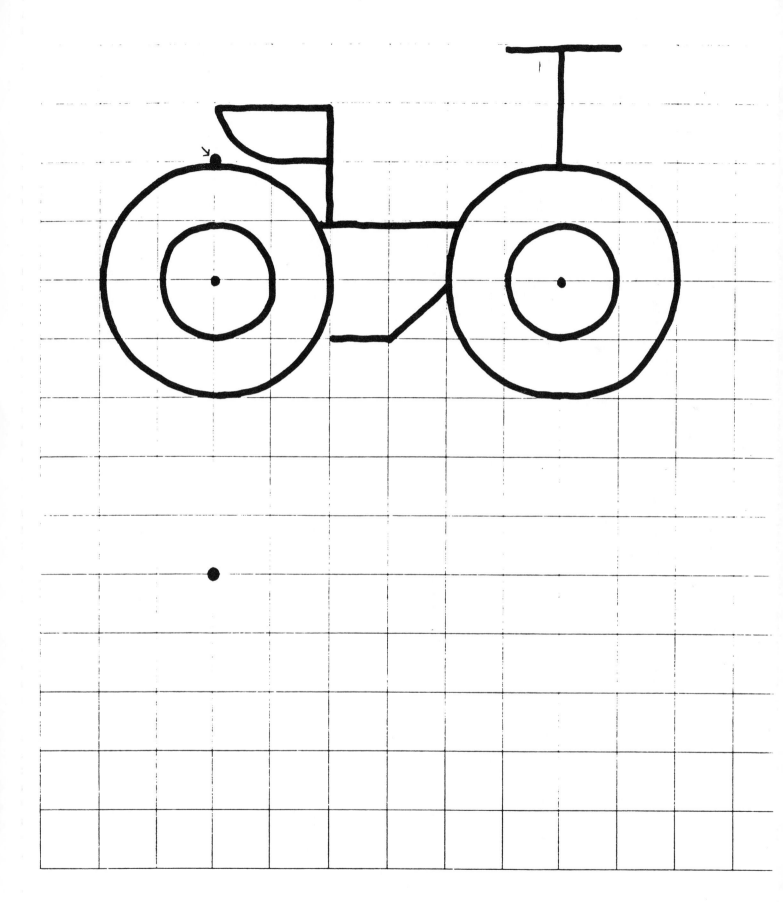

63

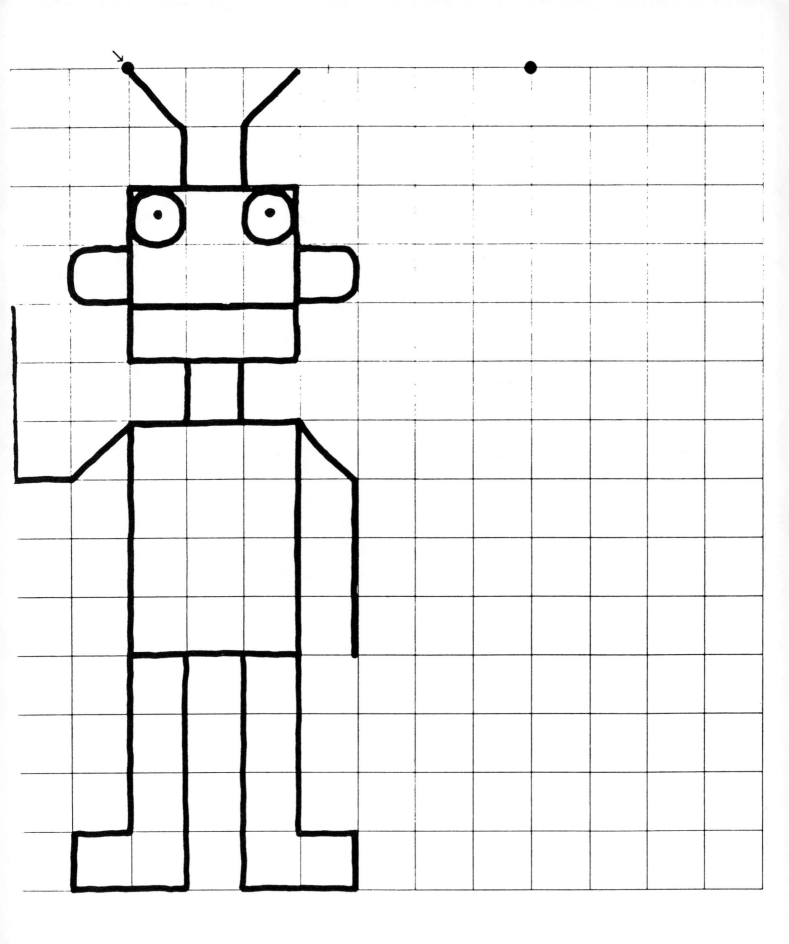

64

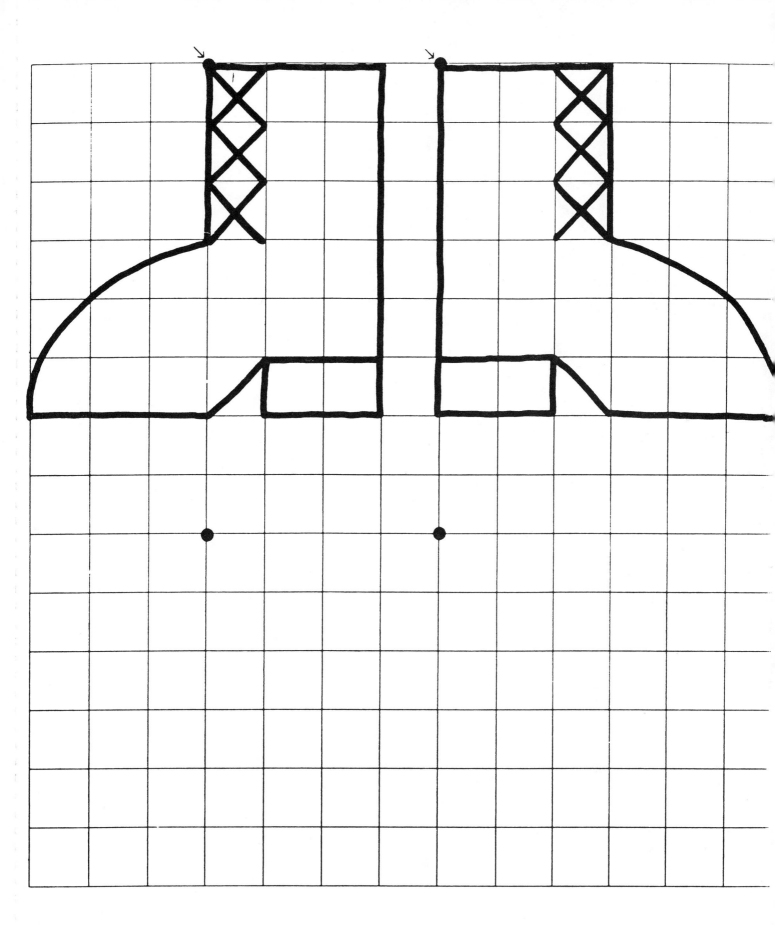

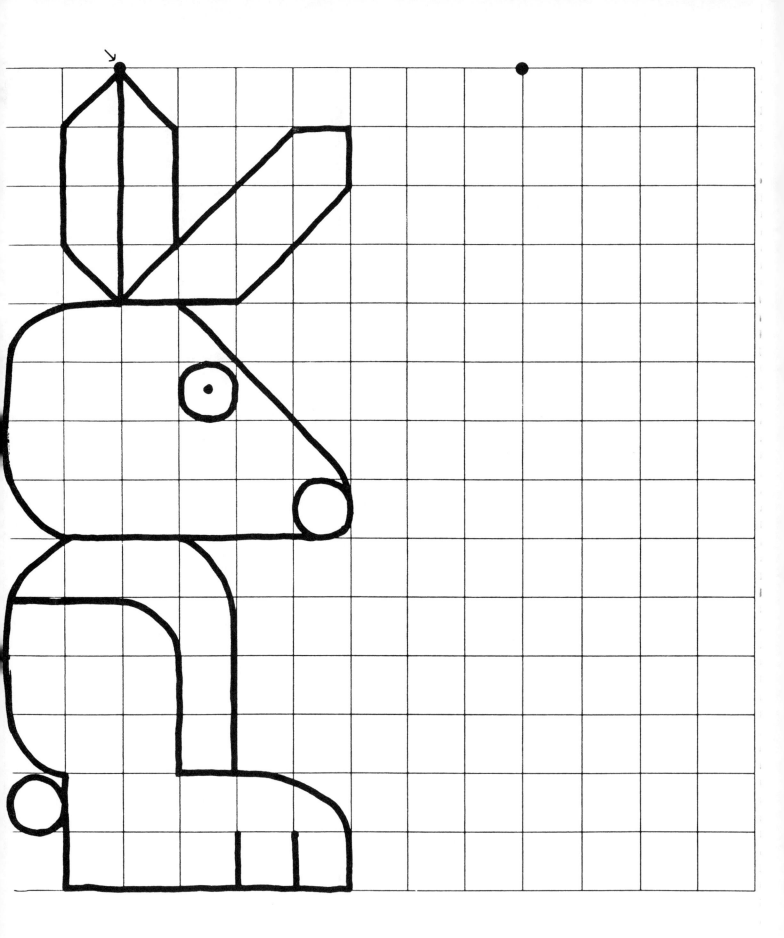

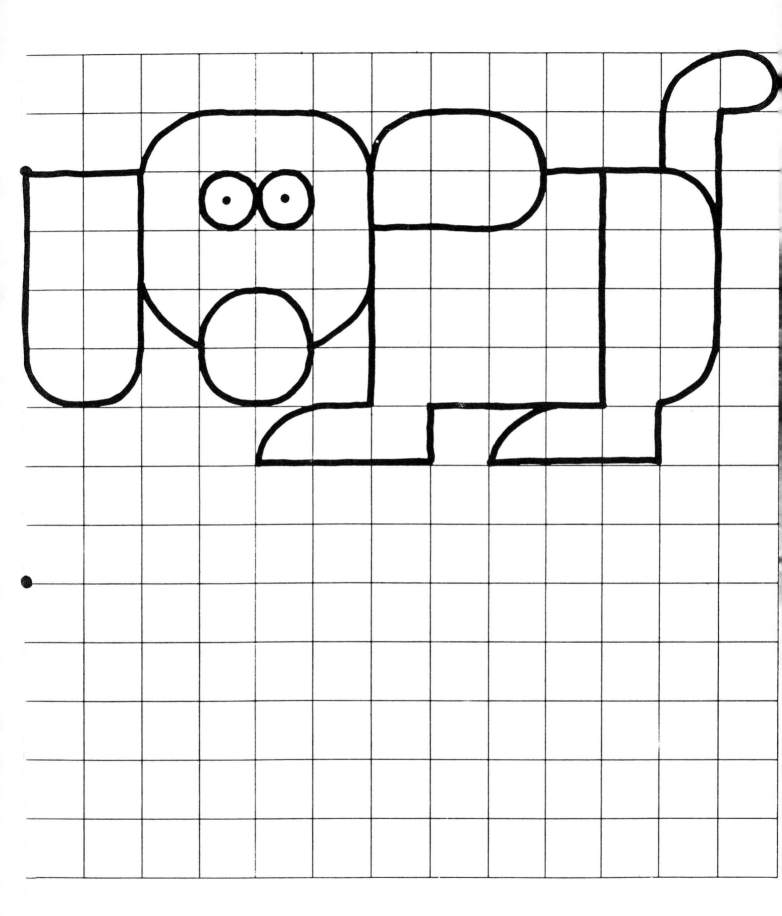

Butterfly Symmetry
March
Grade Level: Primary

TASK ANALYSIS: 2- Relates objects in the environment to geometric figures (squares, circles, triangles, and rectangles)
3 – Folds figures to show symmetry

MATERIALS: Tempra paint, 9" x 12" (22.5 cm x 30 cm) white drawing paper, pipe cleaners, construction paper scraps, paint brushes, glue, direction cards, tagboard pattern

ORGANIZATION: K–3
Station activity with 1 to 2 children working at a time

PROCEDURE: – Direction cards should say:
1. Fold paper in half.
2. Trace tagboard pattern and cut out.
3. Open up and paint on one half of the butterfly only.
4. Fold paper in half again and blot by rubbing hard.
5. Open up and let dry.
6. Add body and antennae.
Children share and discuss the symmetry of their butterflies.

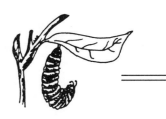

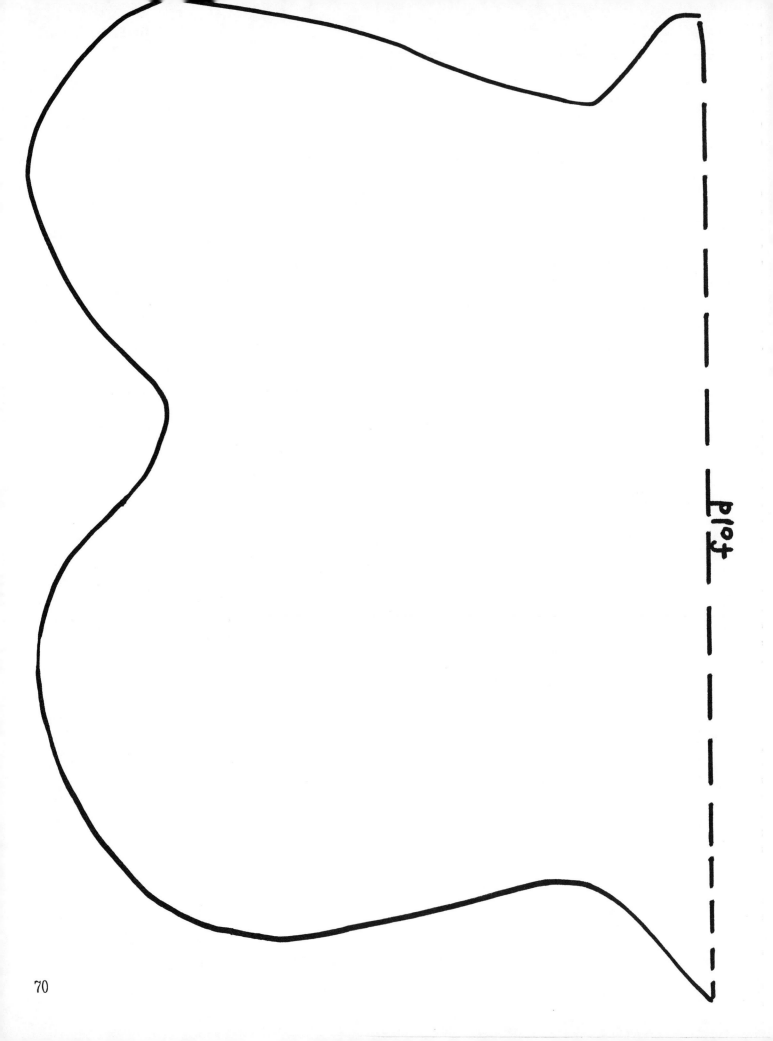

fold

Up and Down Shamrocks
March
Grade Level: Primary

TASK ANALYSIS: 5 – Identifies simple congruent figures

MATERIALS: Tagboard pattern of shamrock, large green construction paper, 2 other colors of green construction paper, glue, scissors, pencils

ORGANIZATION: K–3
Whole class presentation
Station activity

PROCEDURE:
- Teacher models the following procedure:
- Trace and cut shamrocks from 2 colors of green construction paper (example: lime green and dark green).
- Glue one row of (lime green) shamrocks UP on the large piece of green construction paper.
- Glue one row of (dark green) shamrocks DOWN on the green construction paper.
- Continue UP – DOWN rows to the bottom of the construction paper.
- Children follow these steps at the activity station.
- Laminate and use as placemats for St. Patrick's Day party.
- Children discuss and identify the congruent patterns.

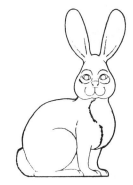

Rabbits, Rabbits, Everywhere
April
Grade Level: Primary

TASK ANALYSIS: 5 – Identifies simple congruent figures

MATERIALS: Rabbit pattern, black and white construction paper, 12″ x 18″ (30 cm x 45 cm) light blue construction paper, scissors, glue, pencil

ORGANIZATION: K–3
Either whole group activity or small group activity

PROCEDURE:
- Child traces rabbit pattern on white paper.
- Child cuts out pattern with white AND black paper together.
- Children decide how many are needed to make rabbit checkerboard.
- Child glues rabbits in checkerboard pattern: 1 white rabbit UP and 1 black rabbit DOWN.
- Discuss.
- These may be laminated and used as placemats.

74

Symm "egg" trical
April
Grade Level: Primary

TASK ANALYSIS: 3 – Folds figures to show symmetry

MATERIALS: 12" x 18" (30 cm x 45 cm) white drawing paper, egg pattern, crayons, pencils, scissors, felt tip pens (optional)

ORGANIZATION: K–3
Teacher directed lesson or a station activity

PROCEDURE:
- Fold the paper in half the long way.
- Trace 1/2 egg pattern on folded paper and cut out.
- Open paper egg and decorate the egg symmetrically.
- See fall symmetry lessons if this is children's first symmetry experience.
- Discuss children's eggs.

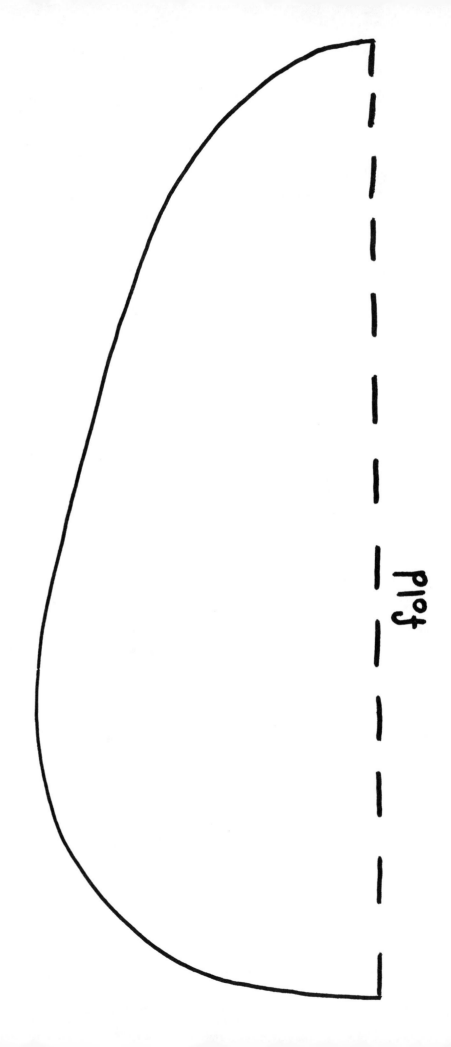

fold

Pattern Block Stories
April
Grade Level: Primary

TASK ANALYSIS: 7 – Use prepositions to describe relationship (location) between two or more items

MATERIALS: Pattern blocks, paper pattern block shapes, white drawing paper, story paper, pencil, glue

ORGANIZATION: K–3: Children working individually

PROCEDURE:
- Children make a story with their pattern blocks.
- Children write a story using prepositions.
- Example: "The yellow hexagon sat on the blue diamond. The blue diamond screamed, 'Ouch!'" "Along came red trapezoid and said, 'Get off, yellow hexagon. You should go and see green triangle.'" "But, where is green triangle?" "He's between orange square and tan rhombus." Students continue story.
- The children will use paper shapes to illustrate their stories.

Pattern Blocks

Elastic Shapes
April
Grade Level: Primary

TASK ANALYSIS: 8 – Identifies open and closed figures

9 – Identifies and names point and line

10 – Identifies and draws horizontal, vertical, diagonal lines

MATERIALS: A ten foot length of thin elastic. Tie ends together to form a circle.

ORGANIZATION: K–3

Whole class activity with children in groups of 4.

We suggest this be an outside activity.

PROCEDURE: – Place children into cooperative groups of four.

– Teacher (or a child) gives direction: "Make a triangle with your elastic circle."

– Children work together to complete task.

– Children then tell how they created the triangle.

– "Make a trapezoid." Follow same procedure.

– Continue until all the geometric shapes have been created.

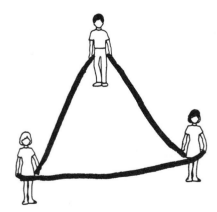

Petals

May

Grade Level: Primary

TASK ANALYSIS: 2 – Relates objects in the environment to geometric figures (squares, circles, triangles, and rectangles)
3 – Folds figures to show symmetry
7 – Use prepositions to describe relationship (location) between two or more items

MATERIALS: Petal pattern, wallpaper or fabric or wrapping paper scraps, colored construction paper, green construction paper for leaf, pipe cleaners, glue, scissors, markers

ORGANIZATION: K–3
Teacher directed or station activity

PROCEDURE: – Discuss how flower gardens are designed (rows, color, height).
– Discuss how many petals are needed to make the flower symmetrical, how many leaves on the stem, etc.
– Trace petal pattern and cut out to create flowers.
– Trace and cut out center of flower.
– Cut out leaves and attach to pipe cleaner stem.
– Combine pieces to create a flower.
– As whole class, "plant" symmetrical bulletin board flower garden.
– Discuss using prepositions.

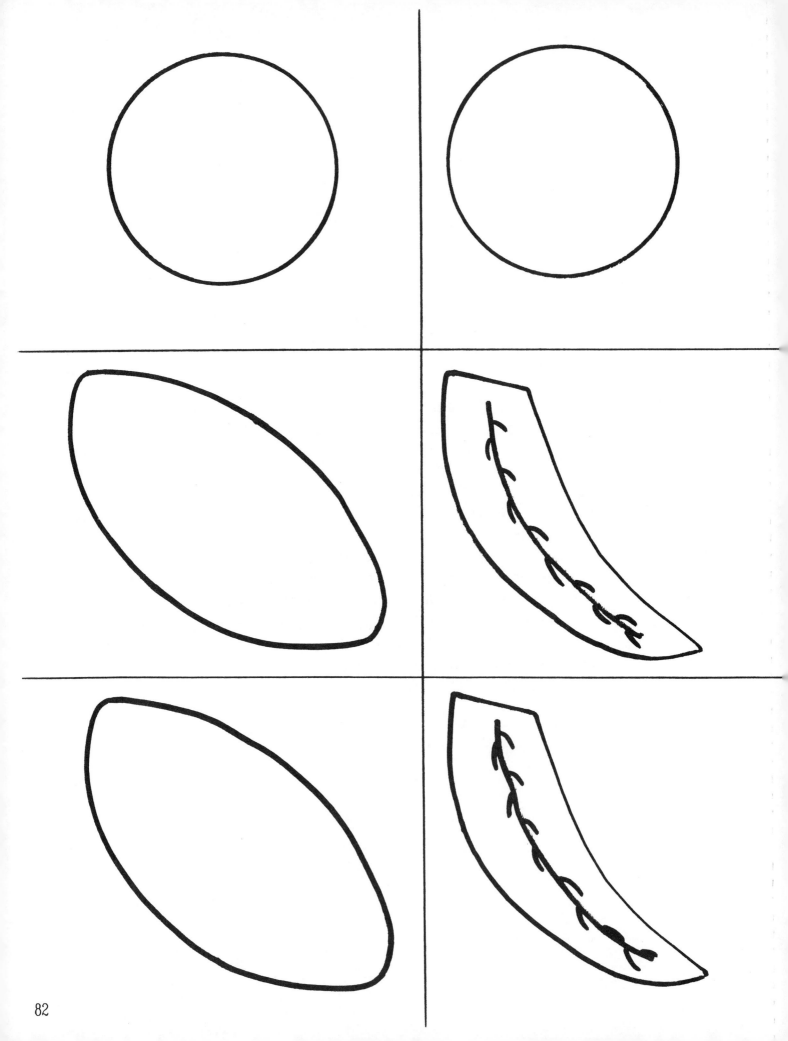

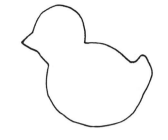

Chick, Chick, Here & There
May
Grade Level: Primary

TASK ANALYSIS: 5 – Identifies simple congruent figures

MATERIALS: 12" x 18" (30 cm x 45 cm) white drawing paper (for a placemat), several chick patterns, unwrapped crayons

ORGANIZATION: K–3
Station activity

PROCEDURE: – Children free explore for the possible positions for placement of the chick pattern around the border of the placemat.
– Children place chick pattern under paper (any position).
– Rub over top of pattern with the flat side of the unwrapped crayon.
– Create a border of congruent chicks.
– Laminate and use as placemats.

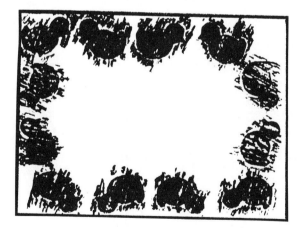

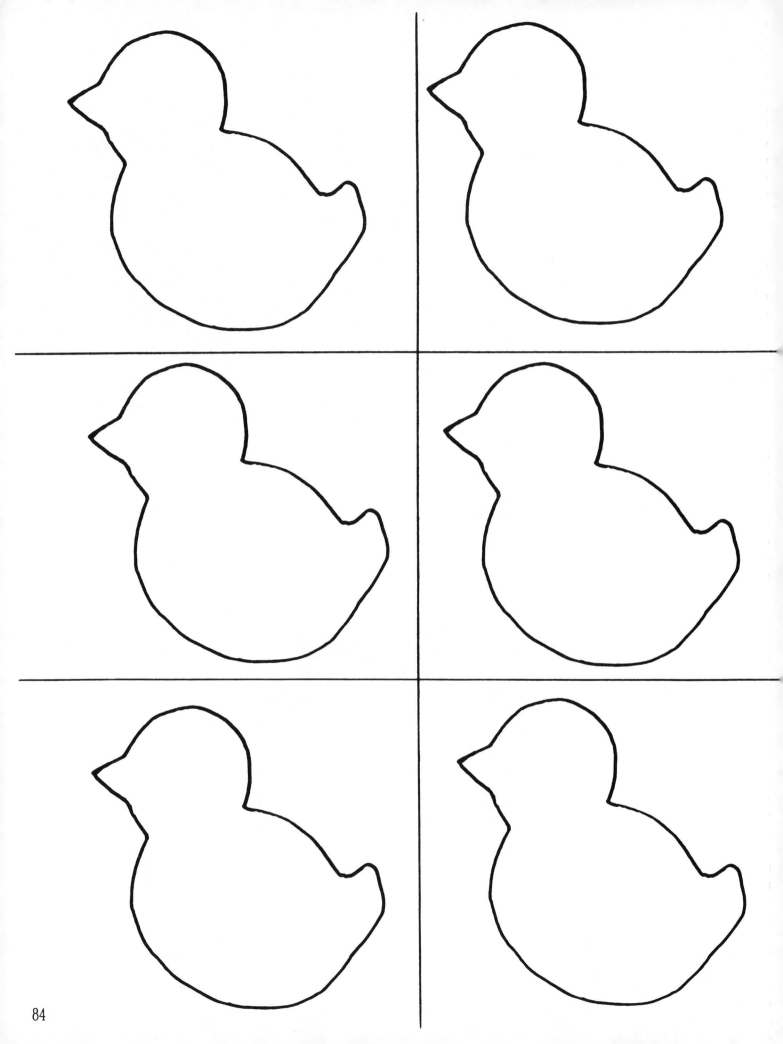

84

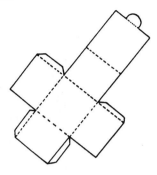

Toy Box Geometry
May
Grade Level: Primary

TASK ANALYSIS: 2 – Relates objects in the environment to geometric figures
(square, circles, triangles, and rectangles)
4 – Matches solid shapes (cube, cone, cylinder, sphere, rectangular prism) to objects
9 – Identifies and names point and line

MATERIALS: Toy box pattern run off on colored construction paper, geometric shapes on colored construction paper, glue, scissors, pencils

ORGANIZATION: K–3
Whole class activity

PROCEDURE: – Brainstorm for ways to help Mother.
– Make toy box (see pattern).
– Teacher models procedure.
– Children choose geometric shapes on which to record "jobs".
– Children cut out shapes.
– Children may write or dictate "jobs" to an adult.
– Children fill the toy chest with their "job" shapes.
– Discuss the geometrical figures, solid shapes and lines used.

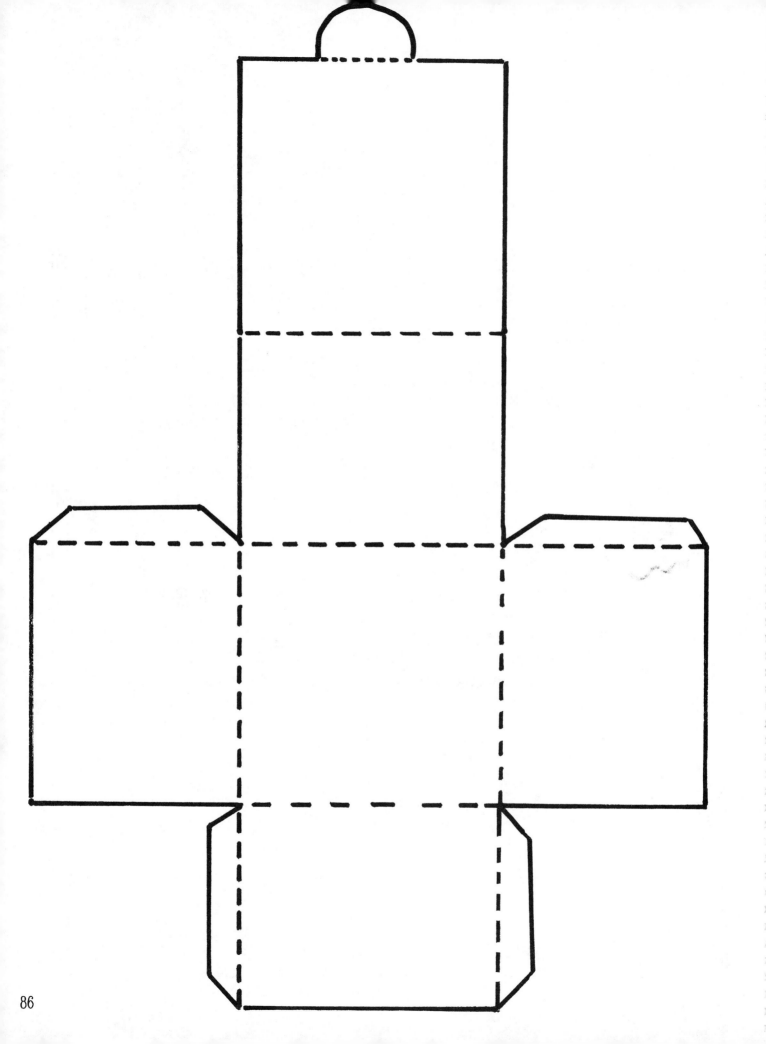

86

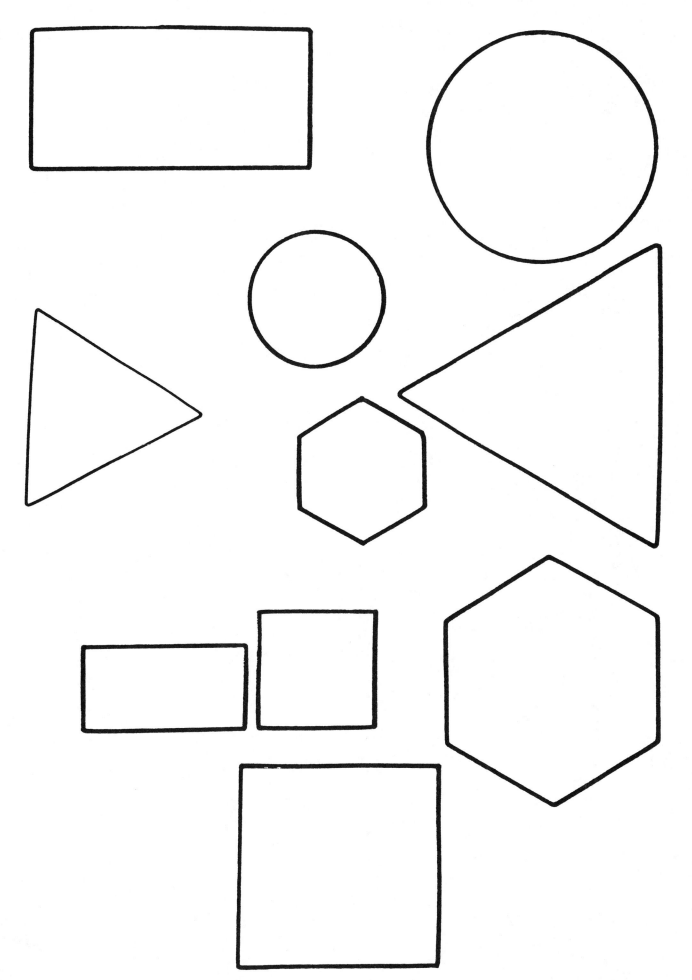

Papa's Necktie
June
Grade Level: Primary

TASK ANALYSIS: 2 – Relates objects in the environment to geometric figures (squares, circles, triangles, and rectangles)
3 – Folds figures to show symmetry

MATERIALS: 12"x 18" white drawing paper for tie, colored construction paper scraps, neckties from home (children bring), tempra; sponges, scissors, glue

ORGANIZATION: K–3
Teacher directed or station activity

PROCEDURE: – Teacher models folding the paper the long way.
– Trace necktie pattern on folded paper.
– Cut out necktie shape.
– Open shape and decorate symmetrically using any available materials.

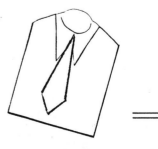

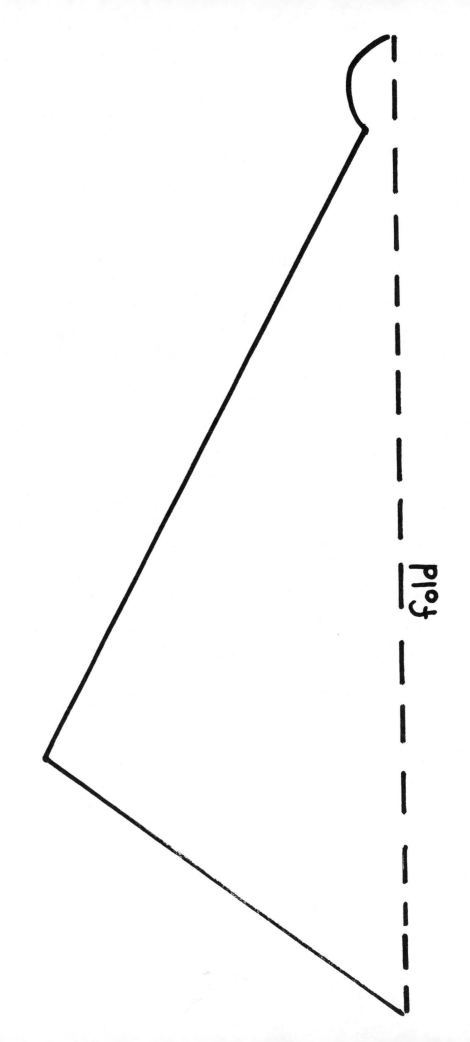

fold

90

Argyle Socks
June
Grade Level: Primary

TASK ANALYSIS: 1 – Compares, contrasts and classifies circles, triangles, squares, rectangles, trapazoids, hexagons, rhombus and diamonds
2 – Relates objects in the environment to geometric figures (squares, circles, triangles, and rectangles)
3 – Folds figures to show symmetry
6 – Recognizes angles (corners)
10 – Identifies horizontal, vertical and diagonal lines and draws

MATERIALS: Argyle socks, paper sock pattern, scissors, pencils, crayons

ORGANIZATION: K–3
Children work in pairs with unmatched pair of socks.
Cards are made individually.

PROCEDURE: – Children bring examples of argyle socks (be sure socks are marked).
– Children working in pairs find horizontal, diagonal, and vertical lines on the socks. Discuss.
– Children find angles on socks. Discuss.
– Teacher models the making of the argyle sock card.
– Children work individually to make the card.

– Extension: Present a challenge! Have the children glue geometric shapes on paper sock to form the argyle pattern.

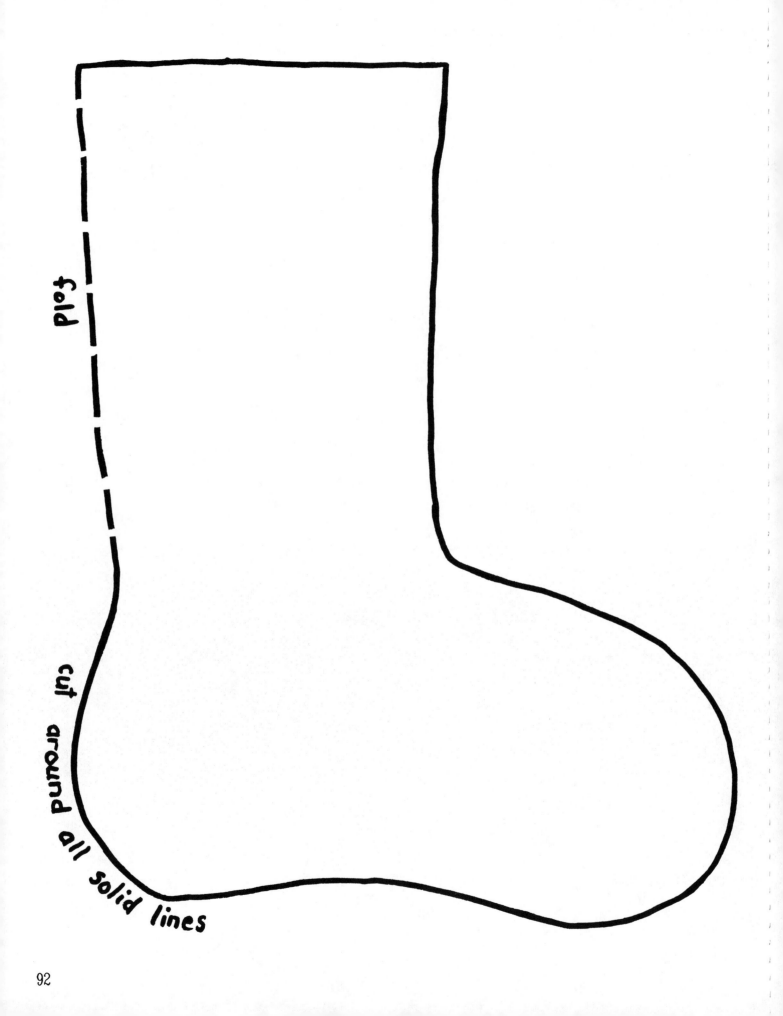

fold

cut around all solid lines

Necktie Laundry
June
Grade Level: Primary

TASK ANALYSIS: 5 – Identifies simple congruent figures

MATERIALS: 12"x 18" white drawing paper, string (to make clothesline), paper clips (clothespins), necktie pattern, scissors, crayons

ORGANIZATION: K–3
Whole class activity for presentation
Station activity with one or two children

PROCEDURE: – Teacher models:
– Fold paper the long way.
– Cut out 1/2 necktie shape (or have children trace pattern).
– Cut out snips from folded necktie shape.
– Open folded necktie to show design.
– Children hang their ties on the "clothesline", to create a congruent pattern.

– Send neckties home for Father's Day.

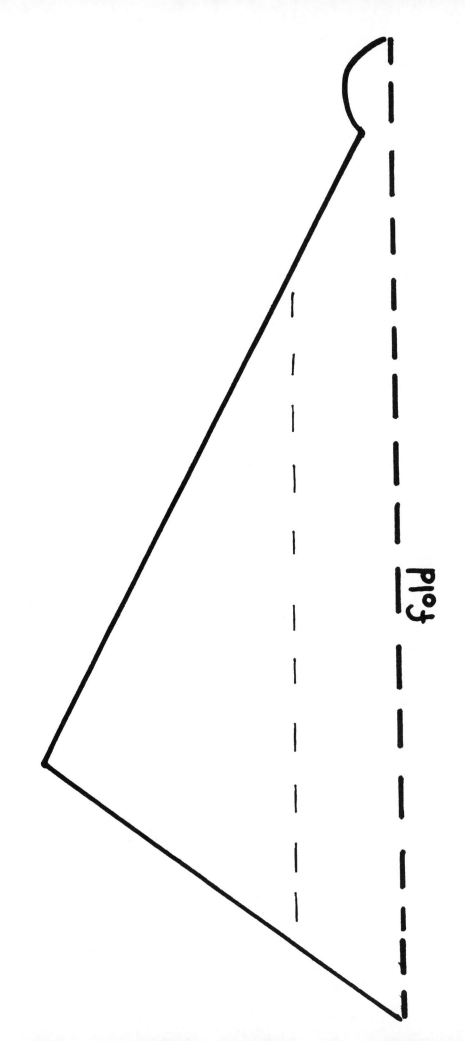

fold

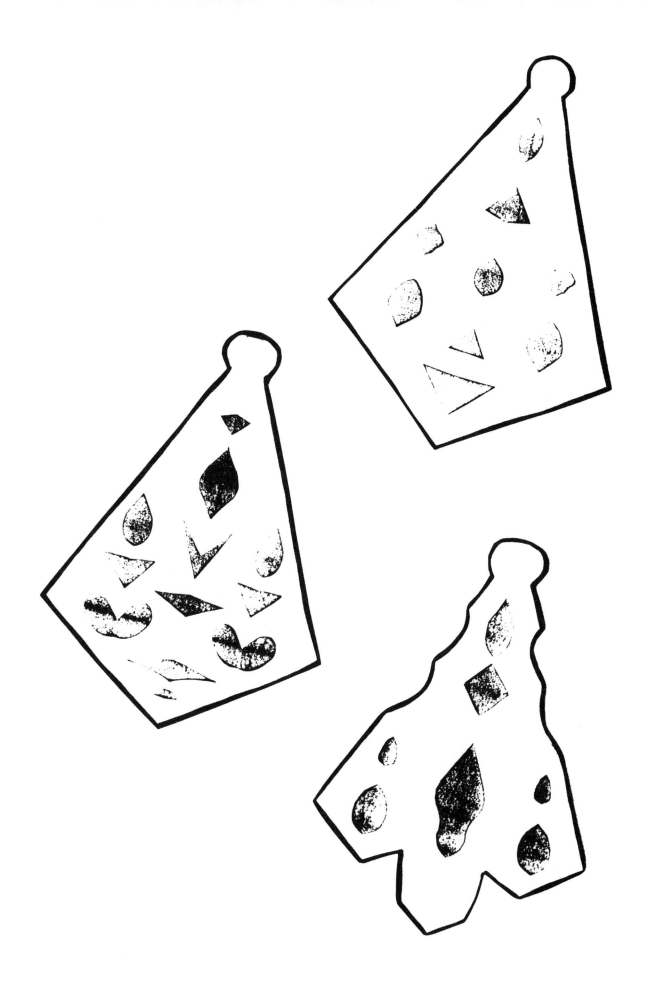

11	Identifies relationships and constructs representations of various one dimensional figures, their symbols, and their properties: point, line, line segment, ray.

Itemize Your Deductions
Grade Level: Middle

MATERIALS: Butcher paper, markers

ORGANIZATION: Cooperative groups of four

PROCEDURE: In this activity, students will be working in cooperative groups. The object of the lesson is for each group to create a table which categorizes various points, lines, line segments, or rays as described by the teacher. Parallel, perpendicular, and intersecting lines may also be included in this lesson.

Begin by giving each team a three to four foot length of construction paper and a marker. Have students make and label a chart as shown. Tell students that for each item the teacher identifies, they must decide if it is a point, line, line segment, or ray. If so, they should place an "x" in that column. Alert students to the fact that certain items may fall into more than one category.

Some samples the teacher might include are: the horizon of the ocean, a flashlight beam, the north pole, the equator, a kite string attached to a flying kite, etc.

These questions will spark discussion within the groups as well as with the entire class when you go over the results.

	Point	Line	Line segment	Ray	
1.		X			horizon of ocean
2.			X	X	flashlight beam
3.	X				north pole
4.		X	X		equator
5.			X		kite string

11	Identifies relationships and constructs representations of various one dimensional figures, their symbols, and their properties: point, line, line segment, ray.

Mapmaker, Mapmaker, Make Me a Map
Grade Level: Middle

MATERIALS: Five different colors of marking pens, tracing paper, street map of your town for each group

ORGANIZATION: Individually or in cooperative groups

PROCEDURE: In this activity students will practice their skills of identifying and constructing one dimensional figures such as points, lines, line segments, and rays. Students may need review of the characteristics of the one dimensional figures mentioned above.

Begin by explaining to the students that they will be tracing portions of their city map in a special way (the map should be in a relatively large scale so students can see its features easily). They will need to create a legend which identifies a color for a given property (i.e. red for points, blue for line segments, green for rays, and orange for lines).

Once this legend is created, the students should place tracing paper over the map and carefully trace each one dimensional figure in the correct color. The vast majority of lines will be line segments, but this analysis is part of the purpose of the lesson. Students will probably ask about other features such as curved lines (arcs) or angles. You may wish to have them add these to the legend.

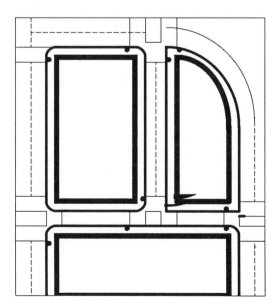

When students have completed the assignment, have them exchange papers with one another and ask questions about the map. The maps will also create a nice bulletin board display.

11	Identifies relationships and constructs representations of various one dimensional figures, their symbols, and their properties: point, line, line segment, ray.

Merry Go Round
Grade Level: Middle

MATERIALS: Sheets of butcher paper, straight edge, markers, small jars of poster paint, or crayons

ORGANIZATION: Groups of four to six students

PROCEDURE: In this merry-go-round activity, students will create a wall mural by drawing one dimensional figures while rotating around the mural.

Begin by dividing the students into groups of four to six. Have them stand around a table covered with a piece of butcher paper. Each student should have crayons, small jars of poster paint, or markers of various colors.

The teacher should then name a one dimensional figure (such as "line segment") and the students should draw the figure on the butcher paper. When students have finished, the teacher says "change," and students rotate one space in a clockwise direction. The teacher then names another figure and students select a color and draw this. Continue this procedure until students have rotated around the table and created a mural.

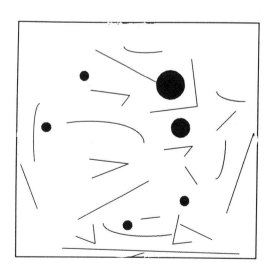

Students may draw the "pictures" any length or any color but the activity works best if students do not intersect lines.

As an extension, have students draw various two dimensional figures.

11	Identifies relationships and constructs representations of various one dimensional figures, their symbols, and their properties: point, line, line segment, ray.

Stories With a Point!
Grade Level: Middle

MATERIALS: No special materials are necessary

ORGANIZATION: Individually or in groups of two to four

PROCEDURE: This activity will give students an opportunity to do creative writing and illustrations using elements listed in the task analysis.

Begin by reviewing the attributes of a line, a line segment, a point, and a ray. Tell students that they are going to write a story in which these one dimensional features are the main characters. You may choose to have students do the writing as an illustrated story, in a picture book format with a picture for each thought or sentence, or a comic book format with the characters speaking to one another.

You may need to provide some motivation by leading students into the assignment. We have provided a sample story for this purpose. Once you have read this story to the children, discuss some different scenarios which they could create, as well as how this story might be illustrated. You will find that some students will use the story line as given here, but many students will take off in other directions to create their own stories that have a point!

"Lines, lines, lines, that's all I ever see," said the ray mumbling to himself as he walked along the city streets. "Why can't the world be made up of something more interesting -- like rays!"

About this time, two very attractive lines came walking by the ray and overheard his mumbling.

"What do you mean you're tired of lines? You shouldn't think of all lines as one group! We happen to be very special lines; we are line segments," replied passerby number one.

"Yes, and there are two points about us which are very important! We have a beginning and we have an end, unlike those lines that you so willingly criticize," added passerby number two.

"You mean there is a difference between a line and a line segment?" questioned the ray.

"Of course," said passerby one, "and now you're beginning to see the point!"

"What does a point have to do with this?" yelled a small little dot, hopping up to the threesome.

"Why you're nothing but a spot," said ray.

Point responded in an all-knowing way, "Yes, I am just a spot -- or dot -- but without me, you would be nothing more than a line. And you line segments would be nothing but lines. You see, you may complain about my size and my looks, but without me, you would have no beginning and no end. That's a point to think about!"

12	Identifies relationships and constructs representations of various one dimensional elements and their symbols: intersection, bisections, parallel, and perpendicular.

Parallel, Perpendicular Collage
Grade Level: Middle

MATERIALS: Magazines, scissors, glue

ORGANIZATION: Individually or in cooperative groups

PROCEDURE: In this activity, students will search through magazines for pictures which show figures parallel or perpendicular to one another.

Begin by reviewing the concepts of perpendicular lines and horizontal lines. Identify several items in the room which are perpendicular to one another or are parallel. Students should eventually come to the conclusion that between two objects there are only three relationships which can exist. They can be perpendicular to one another, they can be parallel, or they can be at an angle. These relationships may seem obvious, but students should recognize this. The collage is an excellent means of grouping this information.

Hand out magazines and scissors to each student or group and let them be creative as to how they want to organize the collage. They might also write a paragraph explaining why they organized the collage as they did. When they have finished, post the collages on a bulletin board and use them for the focus of future class discussions on parallel and perpendicular.

12	Identifies relationships and constructs representations of various one dimensional elements and their symbols: intersection, bisections, parallel, and perpendicular.

Line Finding
Grade Level: Middle

MATERIALS: Magazines and assorted drawings (black and white line drawings are most effective)

ORGANIZATION: Individually or teams of three or four students

PROCEDURE: In this activity, students will analyze drawings for various one dimensional elements and will label these features as they are found.

Begin the lesson by reviewing point, line, line segment, intersection, acute angles, obtuse angles, parallel lines, and perpendicular lines (and any other one dimensional features that you might want to review). On the chalkboard or overhead, create a legend for each of these elements. A sample is shown below.

Next, give each student or group a magazine or drawing and have them analyze the drawing for the various components listed in the legend. They should label these as they find them. Some of the points of lines on the drawing will have two or three labels. This is one of the purposes of this lesson: to show that lines or relationships between lines can be multi-faceted.

Once students have finished the project, have them display their work on a bulletin board.

LEGEND	
A	point
B	line segment
C	intersecting lines
D	acute angle
E	obtuse angle
F	arc
G	parallel lines
H	perpendicular lines

12	Identifies relationships and constructs representations of various one dimensional elements and their symbols: intersection, bisections, parallel, and perpendicular.

1-2 Pick Up Intersections
Grade Level: Middle

MATERIALS: Several sets of pick-up sticks (toothpicks can be used as well)

ORGANIZATION: Cooperative groups of four students

PROCEDURE: This activity is similar to the game of pick-up sticks, with the variation that in this game, students have to identify the relationship between two sticks before they pick them up.

To teach the game, you might wish to bring small groups of children around a table. The game is too small to be seen by the whole class at one time. You might also experiment with explaining the game on the overhead projector.

Have students hold the sticks as a bundle and let them fall randomly. Overlapping is expected. The first student then identifies two sticks by saying, "I'm picking up the two sticks which are intersecting one another (pointing to the sticks)." Other possible identifying terms might be: the sticks which are parallel to one another, the sticks which are at an angle to one another, a single stick which is a line segment, or the sticks which are perpendicular to one another (an added feature might be to include acute and obtuse angles). One rule which you may want to add is players may not identify the same relationship two times in a row.

The student then tries to remove the identified sticks. If anything moves while they are picking up their selected sticks, they must place them back in the pile. If they successfully remove the sticks, they keep them and the next player takes a turn. The student with the most sticks at the end of the game is the winner.

Should they incorrectly identify the relationship between the sticks, they lose their turn and the next player plays.

12	Identifies relationships and constructs representations of various one dimensional elements and their symbols: intersection, bisections, parallel, and perpendicular.

What's Your Angle?
Grade Level: Middle

MATERIALS: Toothpicks, paper, and glue

ORGANIZATION: Individually or in groups of two

PROCEDURE: Begin by a describing a right angle. Tell students that they are going to use three line segments (represented by toothpicks) to create as many different right angle configurations as possible. Hand out toothpicks and give students time to create the four diagrams shown. Ask students to count the number of right angles in each figure.

The next step is to give students a number of toothpicks and let them create shapes using sets of four toothpicks. Once again, they should label the number of right angles per shape. Students should create different shapes, not merely "rotated" shapes with the toothpicks. Have them check one another for original design. This discussion should include the correct vocabulary (perpendicular, etc.).

Have students glue their shapes to a large piece of paper for discussion and display.

13	Identifies relationships and constructs representations among one dimensional elements: coordinate points/axis points on a coordinate graph or grid.

Grids with a Twist
Grade Level: Middle

MATERIALS: Large piece of white plastic material (large enough for a student to reach from corner to corner), slips of paper, and two paper bags

ORGANIZATION: Groups of five students

PROCEDURE: Students should be familiar with coordinate graphing and number pairs. Prior to the lesson a grid should be drawn and numbered on the plastic (blacktop area of the playground or taped off section of the floor can also be used).

This activity is similar to the Twister game with a leader calling out number pairs and telling one player at a time to place a hand or foot on a specified number pair. Rather than using a spinner, we suggest you write numbers on slips of paper and place them in a bag. Write LEFT FOOT, RIGHT FOOT, LEFT HAND, and RIGHT HAND on other sheets of paper and place them in a second bag. The game proceeds as follows:

Select one student to be the "caller." Player one steps up to the grid and the caller selects one card from each bag and calls, "RIGHT FOOT on 4,2." Student 1 place his right foot at the appropriate intersection. Player 2 then steps forward and the caller selects new slips (after returning the old one to the bag) and calls out, "LEFT HAND on 5,1." Continue this process with players three and four (the game should be limited to four players).

Once each player has had a first turn, a second pair is called and the student must move to this location, keeping the original foot (hand) in the first location. Students are disqualified when they fall down.

A variation would be to use hopscotch rules, draw a grid on plastic, paper, or the sidewalk. The students hop from various number pair points.

13	Identifies relationships and constructs representations among one dimensional elements: coordinate points/axis points on a coordinate graph or grid.

Desktop Grids
Grade Level: Middle

MATERIALS: Chalk, beans, and grid paper large enough for beans

ORGANIZATION: Whole class or small groups

PROCEDURE: The teacher should introduce this activity by passing out small numbered grids and asking the students to place beans on the grid as number pairs are called out or written on the chalkboard. Once students understand how to place beans on this small grid, then they are ready to play a game with a large grid.

Have students arrange their desks into straight rows and columns, if they are not already arranged in this fashion. Be certain they understand the difference between rows and columns and which number is named first in the pair. Call out number pairs and have the students name the person sitting at that point on the classroom grid.

As an extension, draw a large grid on the playground blacktop. Have students stand at various points. Call out the number pair and have the class name the student standing on that point.

Both of these activities can be reversed to calling out the name of the student and having him/her state the number pair of the point at which he/she is sitting or standing.

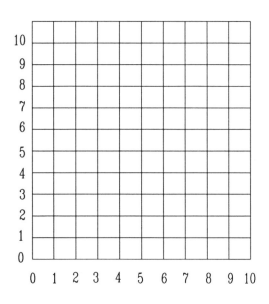

13	Identifies relationships and constructs representations among one dimensional elements: coordinate points/axis points on a coordinate graph or grid.

Stars in Your Eyes
Grade Level: Middle

MATERIALS: Pictures of constellations, graph paper

ORGANIZATION: Individually or in teams of two

PROCEDURE: Young students are usually fascinated with the planets, constellations, and space. This lesson uses their interest to reinforce the concept of coordinate grids.

Begin by discussing the constellations and providing some background history of how each set of stars got its name (there are many picture books which provide this information).

Tell students that they are going to reproduce a constellation on their graph paper. They will then write the coordinate points for the "stars" of their constellation on a separate piece of paper. A coordinate grid for the Big Dipper might look like this diagram.

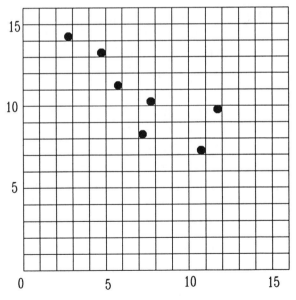

Once students have finished plotting and listing coordinate points, have students exchange their coordinate points papers to see if they can guess the constellation drawn by various classmates.

This lesson will provide practice in identifying the constellations as well as the basic objective of graphing coordinate points on a grid.

13	Identifies relationships and constructs representations among one dimensional elements: coordinate points/axis points on a coordinate graph or grid.

Playing Your Cards Right
Grade Level: Middle

MATERIALS: 1 cm or 1/4" graph paper, four decks of cards (face cards removed), crayons or markers

ORGANIZATION: Groups of four

PROCEDURE: In this activity students will be working with cards to create ordered pairs.

Begin by having your students draw two lines on their paper. The first line will go from top to bottom and be labelled the Y axis (explain to them that the "Y axis" in geometry refers to a line which is used to mark things which are vertical--or top to bottom), the second line will go from left to right and is the X (or horizontal) axis. These two lines should bisect in the center of their papers.

Number the axes in increments of five and then show the class two sample cards. They are to find the location of the POINT on their grid represented by these two cards. The first card displayed will be the X coordinate and the second will be the Y number coordinate. Once this is done, draw a second pair of cards and have students plot a second point. Draw a line through these two points.

Repeat this process to make a second line in the grid.

The challenge for students is to predict the grid location where these two lines intersect, or where they will intersect if extended. Some lines will intersect off the grid and students should identify this (advanced students might want to make predictions about possible locations). Many combinations will create an intersection within the grid.

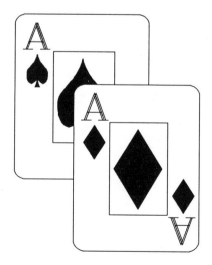

After you have modeled this process a few times, have the students work in teams of four. Each person is dealt four cards. They must plot the two points represented by the four cards and then indentify the point of intersection.

| 14 | Identifies and labels angles (and vertex) according to three lettered points. |

Stick to the Angles
Grade Level: Middle

MATERIALS: Chalk, pencils, paper, various colored pens, stickers or tape, and index cards

ORGANIZATION: Whole class activity

PROCEDURE: This is a very basic beginning lesson on labeling angles. The students would enjoy doing this lesson outdoors.

Assign each child a set of three letter combinations (i.e. ABC, XYZ). Have students find one angle in the classroom and using chalk, index cards, or stickers, have them label the angle. Give them a time limit in which to do this.

After all the labeling has been done, assemble the class. Tell them they are to find each others' angles, record them, and list the vertices. Tell them they will be timed. They may not start until you give the "go" signal. After they are finished, have the students share their results with one another.

Take the children outside and have them repeat the process of finding and labeling angles on play equipment or buildings.

An interesting approach to finding angles other than right angles is for the teacher to stack books at different angles and to allow the students to label them, name the end points, vertices, and identify type of angle. Again, a time limit adds to the excitement.

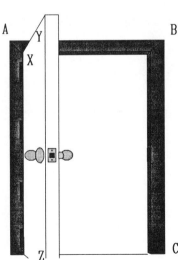

14	Identifies and labels angles (and vertex) according to three lettered points.

The Chair Recognizes...
Grade Level: Middle

MATERIALS: Alphabet letters on cards and a group of chairs assembled in a 5 x 5 square, copies of a grid for each student

ORGANIZATION: Whole class activity

PROCEDURE: Organize chairs in a square with a lettered card on each starting with A in the upper left-hand corner and moving left to right.

```
A  B  C  D  E
F  G  H  I  J
K  L  M  N  O
P  Q  R  S  T
U  V  W  X  Y
```

In this activity students will be practicing their skill in identifying angles. Some students will be sitting in the chairs in the grid with letters in their laps. The teacher should model the activity by calling out a set of three letters, and the corresponding students should stand up. The standing students will have formed an angle. Have the class name the angle, using the correct letter as its vertex.

After you have modeled this procedure, have students create angles on their grids for the seated students to form.

An extension might be to do the same procedure with two-dimensional figures.

| 14 | Identifies and labels angles (and vertex) according to three lettered points. |

Just Stringin' Along
Grade Level: Middle

MATERIALS: Three tennis balls connected by lengths of string

ORGANIZATION: Whole class activity

PROCEDURE: Begin by constructing the "angle string" by punching holes in the tennis balls and feeding the string through. This can be done with any sharp object. Tie knots in the ends, but let the middle/vertex ball slide.

Tell the class that they will work with angles using the newly created "angle string." You may want to review various information about angles before beginning. With the class seated, tell them that you are going to randomly throw the "angle string" out to the class. They are not to jump up or obstruct others from catching a ball. Once each ball has been caught, ask students to identify the vertex and endpoints, also ask them to identify what type of angle it is (obtuse, acute, etc.).

After you have modeled how to toss out the "angle string," have a member of the class throw it. It may take some direction to ensure that everyone gets a chance to participate since the string may not stretch far enough to reach everyone.

An extension might be to have students make estimates as to the size of the angle, or how they could change an obtuse to an acute angle.

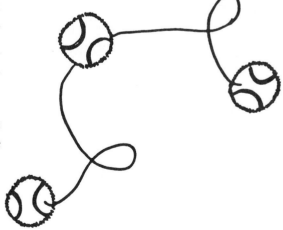

14	Identifies and labels angles (and vertex) according to three lettered points.

Vertex, With End Points Rising
Grade Level: Middle

MATERIALS: String, large tagboard cards with one alphabet letter on each

ORGANIZATION: Groups of three

PROCEDURE: Divide the class into groups of three and then assign each group a 3 letter combination such as ABC or XYZ. Each group should be assigned a different combination. The basis of this lesson is to have various groups come to the front of the room and form different angles for the class.

When a group "performs," each student should hold a card with his letter on it, and the group should hold a string or yarn to show the angle. One student is the vertex and the other two are the endpoints.

To begin, the group may say in unison, "We are angle ABC. Our vertex is B, what are our endpoints?" Or, "We are angle ABC, what is our vertex?" The class should answer with the correct letter. Each group takes turns forming angles and quizzing their classmates.

After the class has become proficient, the performing group should just stand at the front, form the angle, and ask the class the questions without naming themselves. The class should name the whole angle, then the endpoints, and the vertex. The groups can also form shapes using different types of angles and have the class name the type of angle formed, such as acute, etc.

| 15 | Identifies properties of angles: right, acute, and obtuse |

Simon Says "Angle Arms"
Grade Level: Middle

MATERIALS: No special materials are necessary

ORGANIZATION: Whole class activity

PROCEDURE: The object of this lesson is to identify, without measuring, right, obtuse, and acute angles so there is instant recognition of each kind.

To begin the lesson, the teacher should explain the concept of these three angle types. Tell students that they will use their arms (one or two) to demonstrate different angles called out by the teacher.

Have all students stand up. Now begin by making a statement such as, "Simon says, form an acute angle with your right arm." Students who do this action successfully remain in the game. Those who form a right or obtuse angle must sit down.

Continue the game using the variations of right, left, and both arms; acute, obtuse, and right angles; and of course, varying the use of "Simon Says" as you typically would in such a game.

The benefit of this game is that it internalizes the concept of angles for students. They realize that their bodies form angles and this helps them to realize that geometry is not an abstract concept that occurs only in math books.

As an extension, you can elaborate on the game by having students estimate the size of various angles or including vocabulary such as 180° angle and 90° angle. Sample statements might include: "Form a 45° angle with your right arm," or, "Form a 180° angle with both arms."

| 15 | Identifies properties of angles: right, acute, and obtuse |

Stick to the Angles
Grade Level: Middle

MATERIALS: Set of pick-up sticks (16 is a good amount) or an appropriate substitute (pencils, card-board sticks, etc.)

ORGANIZATION: Cooperative groups of three or four

PROCEDURE: In this activity students will be practicing their ability to identify the various types of angles.

The students will need to clear off their desks or find some space on the floor. Have them drop the sticks so that they fall in a random fashion and form a display involving various angles. Have each student create a chart on their paper with three columns, one for each type of angle. After they have dropped the sticks, have them tally the number of each type of angle they can recognize. The person with the highest total number of angles gets to drop the sticks next, or pass to a friend. Before doing so, the winner must identify the angles to the other members of the group.

An extension might be to identify two-dimensional figures using the same method.

15	Identifies properties of angles: right, acute, and obtuse

Angling in on Circles
Grade Level: Middle

MATERIALS: Chalk and a large play area

ORGANIZATION: A whole class activity or divide the class into two or three groups

PROCEDURE: This activity can be done as a P.E. activity in which children practice the concept of acute, obtuse, and right angles (you may also wish to include 180^0 angles as well).

Begin the lesson in the classroom by drawing the diagram as pictured. Demonstrate to students that point V can be viewed as the vertex of many angles (i.e. $\angle AVB$, $\angle GVE$, etc.).

Question the class about the angles which could be formed from the diagram which would be acute angles, obtuse angles, right angles, and 180^0 angles. Some sample questions might include: "If C is an endpoint and V is the vertex, what endpoints would create an acute angle? (B and D). If F is an endpoint, what points would form a right angle? (D and H)."

Once students understand this concept, take them outside and select nine students to be the various points. The game is similar to duck, duck, goose or drop the handkerchief. The teacher should begin the commands, but later, students can perform this role. A sample command would be: "H is the endpoint, V is the vertex, the description is OBTUSE." At this point, the students standing at points E and C would run clockwise around the circle. The first student back to his spot is the winner, the other student is replaced by a classmate waiting in line (since only nine students play at a time, there will have to be a waiting line).

Part of the fun of the game is to watch students when all of a sudden they realize that they are the ones who should be running and their opponent has already begun a trek around the circle.

As an extension, you can play this game with angles (45^0, 90^0, etc.).

15	Identifies properties of angles: right, acute, and obtuse

The Angle in Question
Grade Level: Middle

MATERIALS: Butcher paper, (chalkboard)

ORGANIZATION: Cooperative groups of four students

PROCEDURE: This activity is similar to the "Twenty Questions" game. The variation is that in this game, students have to identify a specific right, acute, or obtuse angle which can be seen in the classroom. Students will first have to discover the angle type and then the location of the angle in order to win.

Begin by dividing the class into cooperative groups of four. Have each group write down the location of an angle. They should keep this secret from the other groups.

One by one, the groups come to the front of the room and field questions from the other groups in their attempt to identify the "hidden" angle." Sample questions might include: "Would running the bases in baseball be similar to this angle?" (right angle) "Would the angle made by opening a desk lid be similar to this angle?" (acute angle) "Would the angle formed by the hood and windshield of a car be similar to this angle?" (obtuse angle).

While these questions are being asked, the teacher should act as a scorekeeper, awarding points to groups for their creativity in thinking of angles which appear in their environment. Points should also be awarded to the group which correctly identifies the location of the angle.

Allow each group on opportunity to share their secret angle with their classmates.

16	Identifies properties of various triangles: base and height, scalene, isosceles, and equilateral

All About Triangles
Grade Level: Middle

MATERIALS: Construction paper and/or cardboard

ORGANIZATION: Individually or in groups of three or four students

PROCEDURE: This is a beginning lesson on properties of triangles.

Run off triangle shapes as shown below. Have the students trace these on construction paper or cardboard and cut them out. This should be done carefully and neatly. The construction paper triangles can be glued to the cardboard triangles or the cardboard triangles can be colored with paint or felt pens. Laminating them would make them more durable for future use.

Some games that might be played include:

1. The students can take turns drawing a triangle from a stack, getting up in front of the class and naming the properties.

2. The students can take turns drawing a triangle from a stack, writing a riddle about it, and asking the class to guess it.

  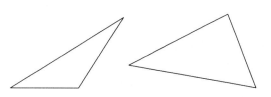 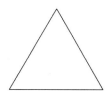

16	Identifies properties of various triangles: base and height, scalene, isosceles, and equilateral

Spider Man Strikes Back
Grade Level: Middle

MATERIALS: Ball of string (possibly two)

ORGANIZATION: Whole class activity

PROCEDURE: In this activity students will practice their skills of identifying the different types of triangles.

Students should stay in their seats for this activity. Have them review the characteristics of the various triangles before starting. When you are ready to begin, bring out the ball of string and tie one end loosely onto one of your fingers. Now unravel some of the string and, with eyes closed, gently lob the ball of string to one of the students. It may take some directing to keep the students from jumping up or interfering with others while they attempt to catch the string. As each student receives the string they should loosely wrap a loop around one of their fingers. Students who have already caught the string should pass or toss the string to someone who has not.

Once everyone is tied together, collect the ball of string and ask students to raise their hand if they are part of a triangle. Have the students who are part of a triangle identify what type of triangle (isosceles, equilateral, or scalene), and who are the corresponding vertices. Ask who would be the base vertices if they were the height or hypotenuse.

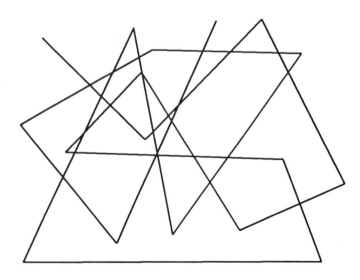

Those students who are not part of a triangle may be asked about those who are, or who they would need to be connected to in order to form a scalene triangle, etc.

16	Identifies properties of various triangles: base and height, scalene, isosceles, and equilateral

Yarn It All!
Grade Level: Middle

MATERIALS: Glue, large sheets of construction paper, scissors, and yarn

ORGANIZATION: Individually or in pairs

PROCEDURE: The purpose of this activity is to give students practice in recognizing and appropriately identifying the three types of triangles. It will also show students how something that is apparently chaotic and shapeless can be given form and definition.

Lay the sheets of paper on a table. Next, take a length of yarn and coat it with glue. Stretch the string in a straight line and place it (pointing in any direction) across the paper. After modeling this for the students, have them work in small cooperative groups to lay yarn segments randomly across the paper and glue them down. Tell students that no more than three line segments of yarn may intersect any given point. This should help promote random distribution. Depending on the size of the paper, everyone should glue at least six line segments of yarn on their paper "planes."

The students' task will be to label the points of intersection alphabetically, starting in the upper left-hand corner and moving across the plane to the right as if reading a comic. Once all the points have been identified, they must then create a chart with three columns: isosceles, equilateral, and scalene. They should now identify all the triangles on their paper and enter them in the correct column on their chart.

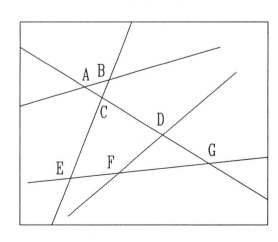

The final step is for the students to take three crayons and color code their triangles.

16	Identifies properties of various triangles: base and height, scalene, isosceles, and equilateral

Triangle Spins
Grade Level: Middle

MATERIALS: Spinner (appendix D), game card (shown below), crayons, paper, ruler and pencil

ORGANIZATION: Small groups

PROCEDURE: Begin by having students identify three types of triangles (isosceles, scalene, and equilateral) and their specific attributes.

Explain that this game is to be played in the following manner:
Remember that the sum of the three angles in a triangle must equal 180^0.
1. Spin two times.
2. Decide if a triangle can be formed (if the total of the two spins is more than 180^0, a triangle cannot be formed).
3. Decide what the third angle would need to be to total 180^0.
4. Decide upon the type of triangle formed by these three angles and color one of the boxes under that heading on the game card.
5. If the total is more than 180^0, or if the three squares for the triangle are colored, the student must wait for his next turn.

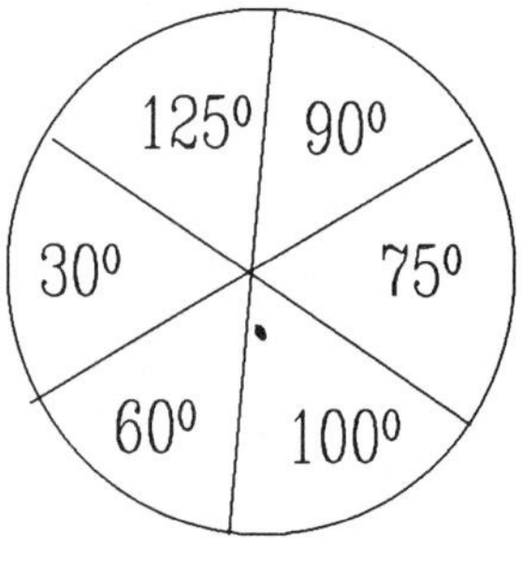

equilateral	isosceles	scalene

GAME CARD

As an extension, you might have students actually draw the triangles with a protractor and straight edge.

17	Identifies properties of various polygons: parallelograms, trapezoids, rhombuses, squares, and rectangles

Analyzing Origami
Grade Level: Middle

MATERIALS: Origami paper-folding books, paper

ORGANIZATION: Individually

PROCEDURE: Most students are familiar with the paper folding art of origami. In this lesson, they will take this art form one step further by being asked to identify each geometric shape they create as they fold their papers.

Begin with an easy folding task and have students follow your directions step by step. With each fold you make (and students make), ask a class member to look at the shape and identify it for the class. As the folds become more complex, students will need to respond with more than one shape. For example, a rectangle, a small and a large triangle.

Eventually, you may want to divide the class into cooperative groups to do group projects of folding and identifying shapes. This lesson also provides practice in following directions.

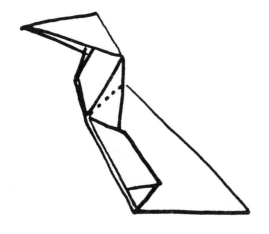 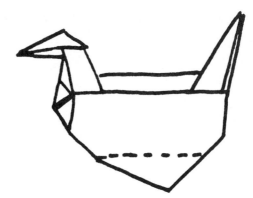

17	Identifies properties of various polygons: parallelograms, trapezoids, rhombuses, squares, and rectangles

Shape Survey
Grade Level: Middle

MATERIALS: No special materials are needed

ORGANIZATION: Individually as a homework assignment

PROCEDURE: This is an early lesson in identifying shapes and finding them in the child's environment.

Begin by reviewing a variety of shapes with the students. Depending upon their sophistication, you may want them to identify three to five different shapes in their rooms at home. Practice this activity by asking students to list examples of all of the circles in the classroom. Discuss their answers and continue this procedure with two or three other shapes.

The students' activity is to make a shape survey of their room at home and to list all of the circles, squares, rectangles, ovals, parallelograms, triangles, etc. You may want to limit it by stating that they should only list those items which can be seen from a certain vantage point or by having them find five or ten items for each shape.

SQUARES	RECTANGLES	TRAPEZOIDS	PARALLELOGRAMS	RHOMBUSES

17	Identifies properties of various polygons: parallelograms, trapezoids, rhombuses, squares, and rectangles

City Shapes
Grade Level: Middle

MATERIALS: Construction paper, marking pens, rulers or straight edges

ORGANIZATION: Individual or small groups

PROCEDURE: Begin this lesson by brainstorming with your students for features of a good city map. Ideas may include: roads, parks, railroads, rivers, airports, lakes, canals, etc. The teacher may wish to post these features for future reference.

The task for each group is to create a map of a fictitious city using geometric shapes to represent the various features. They will need to make a key for the map. Each group should choose a polygon shape that will be used predominately throughout this assignment. If a group should choose the polygon shape of the octagon, most all map features should be made in the shape of the octagon.

Encourage students to be creative not only with their design but their city's name. A preliminary sketch of their planned map may be needed for some groups. Encourage all groups to include as many "map features" as possible on their creations.

17	Identifies properties of various polygons: parallelograms, trapezoids, rhombuses, squares, and rectangles

Half and Half
Grade Level: Middle

MATERIALS: One cm grid paper, scissors, glue, colored markers

ORGANIZATION: Individually or in small groups

PROCEDURE: Begin this lesson by establishing the meaning of congruency versus the concept of being equal in area.

Give each student or group several 4 by 4 cm grid squares and ask them if they can divide the square into two EQUAL parts. There are a variety of answers, so let students experiment with various solutions before discussing this as a class.

Once students have shared their responses, ask them if their equal halves are CONGRUENT to one another. You may choose to have them cut these shapes to more easily check for congruency. In most cases they will be congruent, but not necessarily in all cases. Have students try to divide their remaining 4 by 4 grids into non-concruent, equal area shapes. This experimentation will help students differentiate between the concepts of equal area and congruent shapes.

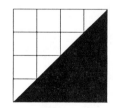

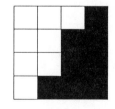
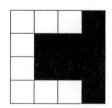

Congruent shape and Equal Area	Congruent shape and Equal Area	Congruent shape and Equal Area	Equal Area but Not Congruent

18	Identifies and labels properties of a circle: diameter, chord, circumference, and radius

Some Total
Grade Level: Middle

MATERIALS: Catalogs, magazines, scissors, glue, paper

ORGANIZATION: Individuals or in small groups

PROCEDURE: Begin by reviewing the elements of a circle (arc, radius, diameter, chord, semi-circle, and circumference).

Tell students that they are going to create collages by cutting out objects that portray different elements of a circle. For example, the lines in a round waffle represent chords, handles on baskets represent semicircles, blades on a fan a diameter or radius, top of a light bulb an arc, etc.

Group the same elements of a circle together in a collage as shown.

Pass the completed collages around from group to group to see if children can identify each area.

18	Identifies and labels properties of a circle: diameter, chord, circumference, and radius

Treasury of Terms
Grade Level: Middle

MATERIALS: Ruler, yard stick or meter stick, string

ORGANIZATION: Groups of three or four

PROCEDURE: Begin by explaining that the groups are going on a treasure hunt for a list of items. Each list will include the elements of a circle (i.e. radius, diameter and circumference).

The children may find the items in their classroom or on the playground. They should measure and record the actual size and location of each item. A time limit should be set. The group of students having the most items within that time limit is the winner. Each group may have an identical list, or the lists may be different for each group.

SUGGESTED LIST:		POSSIBLE STUDENT ANSWERS:
A RADIUS OF:	1–3 centimeters	watch
	1–2 feet	tire
	3–5 feet	tether ball circle
A DIAMETER OF:	11–12 inches	school clock
	1–3 feet	garbage can
	4–10 meters	merry-go-round
A CIRCUMFERENCE OF:	1–3 feet	garbage can lid
	5–10 feet	center of basketball court
	10–20 meters	tether ball court

| 18 | Identifies and labels properties of a circle: diameter, chord, circumference, and radius |

A Blue Plate Special
Grade Level: Middle

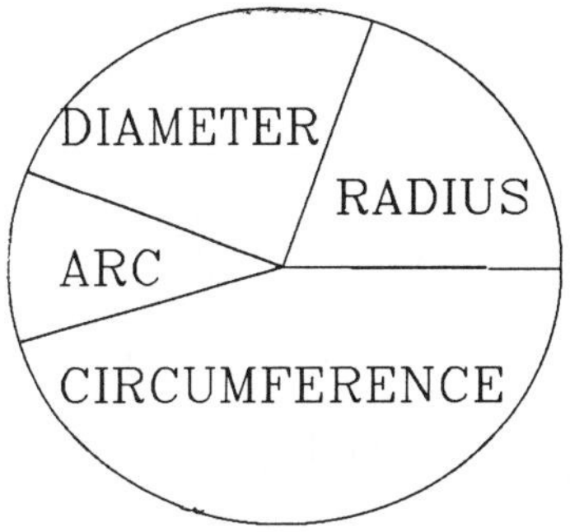

MATERIALS: Paper plates and marking pens

ORGANIZATION: Teams of two students

PROCEDURE: The teacher should begin by reviewing the elements of a circle (diameter, chord, radius, arc, etc,). Tell students that they will be creating a pie (circle) graph on a paper plate. The information contained in the graph will be found by conducting a classroom survey in which students will count the number of diameters, radii, chords and arcs in the room.

Divide the class into teams of two and give each team a paper plate and a marker. Ask them to make a tally sheet similar to that shown below. Give them time to wander about the room finding the various circle parts and have them keep a tally of their discoveries.

After a few minutes, have students return to their seats and explain the concept of a pie or circle graph. In doing this, emphasize the vocabulary of this task analysis. Tell students that they need to try to divide their plates into pieces which represent the number of marks on their tally sheets. For example, if a team found 2 arcs and 6 diameters, the diameter section must be three times larger than the arc section. Also explain that they must use the whole circle so they must plan accordingly.

Give teams time to work out the problem and share their finished products with the class. Discuss why they found more circumferences than arcs; why more diameters than arcs, etc.

DIAMETER	RADIUS	ARC	CIRCUMFERENCE

| 18 | Identifies and labels properties of a circle: diameter, chord, circumference, and radius |

Modern Art Takes Time
Grade level: Middle

MATERIALS: Construction paper, scissors, marking pens, (paint and several modern art sample paintings of Miro, Picasso, Dali, or Jasper Johns optional)

ORGANIZATION: Individually

PROCEDURE: This lesson combines an art project with an exploration of working with circle parts. Students will design a "modern" clock. This clock, however, will serve two purposes, telling time and identifying the properties of a circle. Encourage students to be creative.

Begin by displaying some modern art prints which feature shapes used to create abstractions. Students will usually be able to identify the subject of the art work but will see that by superimposing shapes on one another or by using shapes which are out of proportion, the artists have created abstractions. Students will use these techniques to create their clocks.

Give them the materials listed and have them cut out shapes to be pasted onto a background paper. When their clocks are complete, have them share their creations with the class by pointing out the circle elements they have used.

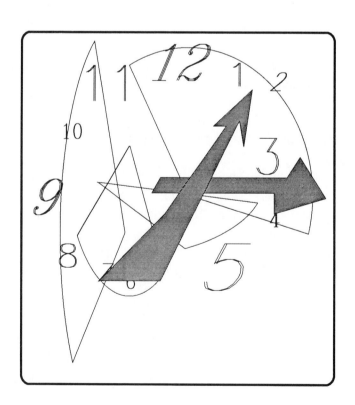

19	Constructs various two dimensional figures: polygons (parallelograms, rhombuses, trapezoids), circles, and triangles (scalene, isosceles, and equilateral)

Letter Perfect
Grade Level: Middle

MATERIALS: Construction paper, scissors, paste

ORGANIZATION: Individually or in groups of two

PROCEDURE: In this activity, students will create their own alphabet script by cutting geometric shapes and gluing them to a piece of paper.

The simple solutions to this problem will involve the use of rectangles and triangles but encourage your students to use a full array of two-dimensional figures. Also encourage students to name or list each shape as it is used.

You can extend this lesson by having students make posters for an art bulletin board with the letters they have created.

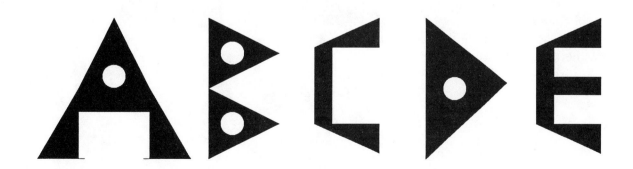

| 19 | Constructs various two dimensional figures: polygons (parallelograms, rhombuses, trapezoids), circles, and triangles (scalene, isosceles, and equilateral) |

A Painting Takes Shape
Grade Level: Middle

MATERIALS: Two landscape art prints, construction paper, scissors, glue

ORGANIZATION: Individually or in teams of two

PROCEDURE: Students need to see that geometric shapes can be found in everything, including art. In this exercise, students will recreate the landscape painting using geometric shapes.

Begin by displaying two landscapes. It is nice to select two different types of scenes and let children select the one they wish to imitate. The landscape should include several different features (e.g., trees, mountains, clouds, lakes, streams, etc).

Ask students to use their imaginations to see if they can see any shapes in the paintings. They should look for squares, triangles, rectangles, circles, and other shapes which you have worked on as a class. Tell the class that their task is to recreate a painting with shapes cut from construction paper and glued on one large sheet. They may use the same colors of the painting, or they may change them if they wish.

As you roam around the room, ask students to name the shapes and encourage them to try some unusual combinations. Once students have finished, these will make a very nice bulletin board display.

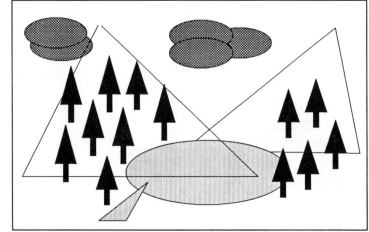

19	Constructs various two dimensional figures: polygons (parallelograms, rhombuses, trapezoids), circles, and triangles (scalene, isosceles, and equilateral)

Decode Is Declue
Grade Level: Middle

MATERIALS: Prepared code sheets (appendix B), yarn, toothpicks, construction paper or tagboard, and glue

ORGANIZATION: Teams of two or small groups

PROCEDURE: The teacher should start this activity with the entire class. Pass out sheets of the code letters (see appendix B) and a word to be decoded. Work with the students to find the letters for the answer.

When the students have decoded the answer, have them make up words or sentences and put them in code. They might do this by gluing yarn and toothpicks on tagboard in the shapes that match the code letters, or the students can simply draw the matching figures. When they have encoded their words or sentences, the codes can be traded so that their team members can decode the secret words.

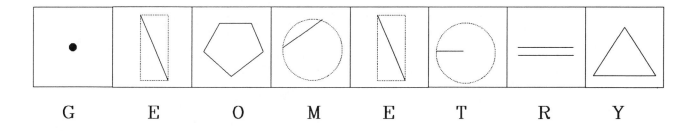

G   E   O   M   E   T   R   Y

19	Constructs various two dimensional figures: polygons (parallelograms, rhombuses, trapezoids), circles, and triangles (scalene, isosceles, and equilateral)

Stay in Shape
Grade Level: Middle

MATERIALS: Magazines, glue, scissors, paper

ORGANIZATION: Groups of two or individually

PROCEDURE: This is an enjoyable lesson for students which borders on nonsense yet teaches basic shapes mentioned in the task analysis. Most students will have heard the phrase "a square meal," but in this case they are going to take the phrase literally. They will also expand the phrase to include a round meal, a triangular meal, or any other shaped meal you would like your class to try to locate.

As a first step, have the students make a poster using magazine cut outs of a nutritional meal titled: A SQUARE MEAL, and A WELL ROUNDED MEAL. The task is to show that each meal is not only well balanced, but in the natural shape of the figure in the title.

For example: A well rounded breakfast might include an egg, an orange, a biscuit, or a pancake. A square meal might include a piece of meatloaf or slice of beef, bread, cubed potatoes, etc. Once students have worked with this concept, have them use their imaginations to come up with other shapes of meals.

By dividing the class into cooperative groups, you will find they will do wonderful posters of geometrically shaped foods for display in your room.

20	Constructs various two dimensional figures to show congruency, ratio, and proportion: polygons, circles, and triangles

Bigger is Better
Grade Level: Middle

MATERIALS: Two ropes (ten to fifteen yards or meters in length)

ORGANIZATION: Whole class P.E. type activity

PROCEDURE: This is a review lesson of the various two dimensional geometric shapes. In this activity, students will work together as a class to move a rope circle into various shapes called out by the teacher and will then try to form proportional shapes using two ropes.

Prepare for the lesson by tying each rope's ends to make two circles. Have each student stand around the outside of one of the ropes and the teacher begins by calling "circle." The students will have to work together to try to form a circle shape.

Continue in the same manner having the children form a square, rectangle, parallelogram, triangle, etc.

Once students have mastered this portion of the lesson, divide the class into two groups and have the newly created group use the second rope. Commands now should indicate proportion between the shapes. For example, the teacher might say, "Group A form a large circle and group B form a circle one-half the size of circle A (for example, double the rope)," or, "Group B form a square and group A form another square three times as large as the Group B square." In the latter case, group B will have to begin by forming a small square so that group A will have sufficient rope to complete their square. This cooperative approach is the key to the lesson.

This activity can also be done in small groups of six to ten children with groups racing each other to see which group can form their shape the fastest.

20	Constructs various two dimensional figures to show congruency, ratio, and proportion: polygons, circles, and triangles

Follow the Leader
Grade Level: Middle

MATERIALS: Geoboards or graph paper, geobands or pencil and ruler

ORGANIZATION: Whole class, then small group activity

PROCEDURE: Begin the lesson by establishing the meaning of the word <u>congruent.</u> Using an overhead projector or the chalkboard, the teacher should demonstrate how to make an isosceles triangle on a geoboard (or graph paper).

Ask students to explain how they could make other congruent triangles by joining other points on their geoboards. After students have had a chance to share their answers, have the class form congruent parallelograms in various rotations or positions on their geoboards.

The next step is to divide the students into small groups and have them take turns being the designer of shapes while others share their congruent imitations in various positions or rotations.

As an extension, you can have students work with various shapes in terms of ratio and proportion.

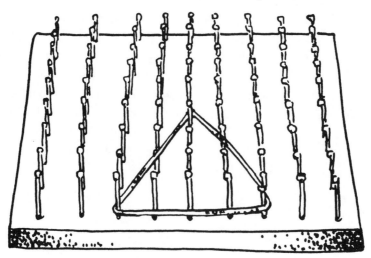

20	Constructs various two dimensional figures to show congruency, ratio, and proportion: polygons, circles, and triangles

Mix and Match
Grade Level: Middle

MATERIALS: Grid paper, scissors

ORGANIZATION: Individually or in small groups

PROCEDURE: Begin by saying that congruent means the same size and the same shape. The teacher explains that the task is to show a set (two, three, or more) of congruent shapes can create a square.

The teacher can demonstrate this by showing that 3 of this shape:

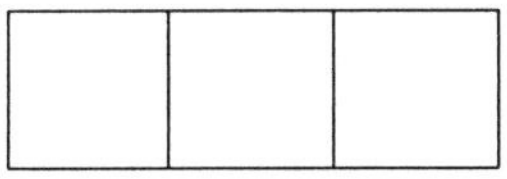

Will create this shape:

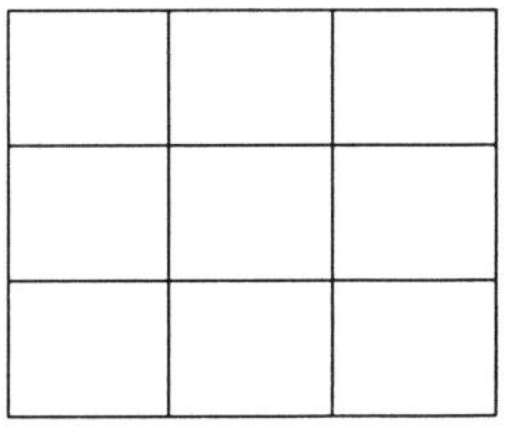

In preparation for students making puzzles of their own shapes, ask them to create a square with four pieces of this shape: and this shape:

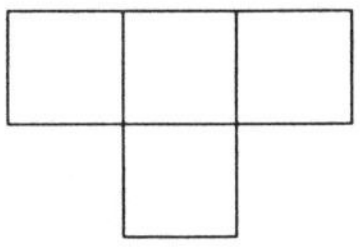

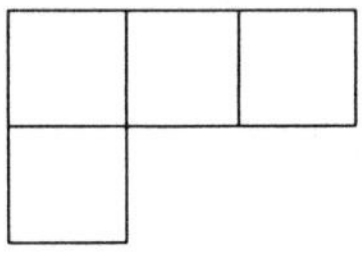

Have students now see if they can create similar puzzles for larger or smaller squares (use of grid paper makes this much easier for students). Exchange puzzles within groups.

20	Constructs various two dimensional figures to show congruency, ratio, and proportion: polygons, circles, and triangles

Look Overhead!
Grade Level: Middle

MATERIALS: Overhead projector, acetate sheets, projector marking pens, large sheets of graph paper or butcher paper

ORGANIZATION: Whole class activity

PROCEDURE: This is a fun lesson in identifying two dimensional figures and dealing with congruency, ratio and proportion among figures. Tell students that they will be predicting and then proving what happens to figures that are expanded (blown up by the projector).

The teacher should begin the lesson by asking students if the figure (a square for example) on the overhead projector will expand or diminish as the projector is moved further away from the graph paper or butcher paper (screen)? Is there any correlation between the inches or centimeters of each move and the relative size of the figure? Discuss and predict.

The teacher then should move the projector back and forth while students measure. Allow students to adjust their predictions.

As an extension, you may wish to have students cut out pieces of paper which estimate the enlarged or decreased size. Let them come to the front of the room and place them over the projected image for comparison purposes.

Later discussions could lead to discoveries of congruent vs. non-congruent figures.

21	Determines lines of symmetry and investigates transformations in two dimensional figures

Stained Glass Shapes
Grade Level: Middle

MATERIALS: Construction paper or colored tissue paper, scissors, glue

ORGANIZATION: Individually or in teams of two

PROCEDURE: This is a review lesson in which students will practice the concept of symmetry and review various types of polygons. Pictures of stained glass windows provide excellent motivation for this lesson.

Give students a variety of colors of tissue or construction paper. Begin by calling out a shape such as, "triangle." At this point, all students should fold a piece of colored paper in half and cut out two identical triangle shapes. Call out another shape, "rectangle," and have students again fold a piece of paper and cut out these shapes. Continue with four or five more shapes.

Next, give students a large sheet of black construction paper and have them lay out their shapes to create a symmetrical stained glass window. Have them check one another for accuracy (you might have them fold the black paper in half first). When this procedure is complete, allow students time to cut out shapes to fill in the spaces on their windows. When the layout is complete, have students glue shapes to the black paper.

If colored tissue paper is used, cut out shapes in the black paper and glue tissue over the open spaces. You can also glue tissue paper shapes on wax paper and outline the shapes with black paper or markers.

21	Determines lines of symmetry and investigates transformations in two dimensional figures

Stationery Action
Grade Level: Middle

MATERIALS: Construction paper, scissors, tracing paper, glue, other materials for creating stationery, American Indian patterns will provide nice samples

ORGANIZATION: Individually

PROCEDURE: This lesson provides an opportunity for students to practice the concepts of symmetry while making personal and original stationery.

Discuss the concept of symmetry with students and tell them that they are going to use a variety of geometric shapes in coordination with symmetrical design to create stationery. This page provides an example for you to share with students.

Students should choose three or four different shapes to make a repetitious pattern for their borders. They should experiment with these patterns before finalizing their stationery.

Depending upon the sophistication and artistic ability of your students, you may need to provide a great deal or very little explanation.

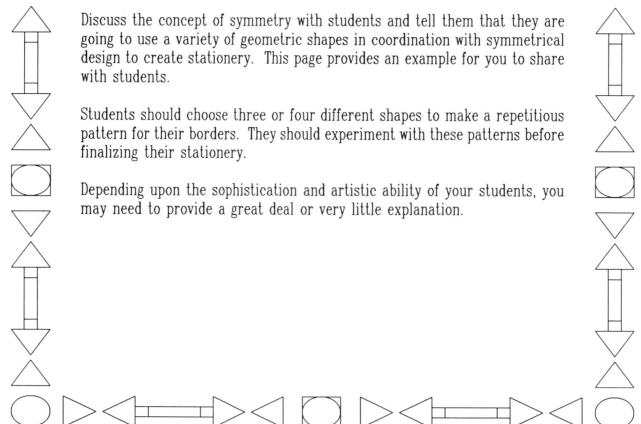

21	Determines lines of symmetry and investigates trans-formations in two dimensional figures

Palindromes
Grade Level: Middle

MATERIALS: Construction paper, scissors

ORGANIZATION: Individually or in teams of two students

PROCEDURE: Most students have experimented with palindromes, that is, words or dates which can be read from either the right or left. Some popular palindromes are 1881, MADAM, and MOM. The ultimate outcome of the lesson is to introduce the concept of symmetry.

By looking at these samples, one can see that the 1881 and madam samples are transverse reversals, but words and dates such as 1961 and mow can be rotated 180° to read the reversal. In this lesson, students will first categorize the letters of the alphabet and the numbers 0 through 9 to establish if they are rotational or transverse. To help them in this endeavor, we suggest you have students cut the letters out of construction paper. Once they have a set of useable letters and numbers, they will create a set of palindromes.

For the teacher's use, the following list includes the various letters of the alphabet and numbers which are transverse and rotational.

Once students have worked with this for a period of time, stop them and discuss the concept of SYMMETRY and explain that the reason that these letters and numbers can create palindromes is because they are symmetrical, that is, the top half and bottom half or the right and left halves are a mirror image of each other.

As an extension, have students go through the lower case alphabet.

Transverse	Rotational
A	H
H	I
I	N
M	O
O	S
T	X
U	Z
V	1
W	8
X	
Y	
1	
8	

21	Determines lines of symmetry and investigates transformations in two dimensional figures

A Fine Line
Grade Level: Middle

MATERIALS: Tracing paper, mirrors, rulers, pencils, newspapers, magazines, or telephone books

ORGANIZATION: Individually or in small groups

PROCEDURE: Tell students that they will be working with various lines of symmetry. If necessary, define the term "symmetry" for children. Explain that in a symmetrical figure, each point on one side of the line of symmetry aligns with a point on the other side of the line of symmetry. You might show some samples.

Tell students that many businesses have a logo which is symmetrical. The reason for this is that symmetry is appealing to people because it is balanced. Given this, have students search through magazines, books, or other materials in search of symmetrical shapes. Have them trace the shapes with tracing paper and then draw the "line of symmetry" on each logo or design.

Have students share their drawings when they are done.

You might extend this activity to have students cut out various symmetrical logos and do a bulletin board of symmetrical shapes.

22	Identifies various three dimensional figures (prisms, cubes, spheres, cones, cylinders, polyhedra, etc.)

Old Mother Hubbard
Grade Level: Middle

MATERIALS: Charts for each student as shown below

ORGANIZATION: Individually as a homework assignment

PROCEDURE: One of the most interesting places to find a variety of shapes is in a student's pantry at home. Different types of cereals, canned vegetables, packages of noodles, etc. can help students to see that shapes they learn about in class are everywhere.

You might begin this lesson by bringing some sample food items from home. If you hold up a cereal box of Oat Puffs, students can identify that the box is a rectangular prism and the Oat Puffs themselves are spherical. Spaghetti noodles would be cylinders, ice cream cones, cones, etc.

Duplicate a copy of the chart below for each student and tell them that their assignment for the evening is to look in the pantry of their kitchens and to identify the shapes of fifteen to twenty items. These shapes should be listed on the chart.

cylinder	rect. prism	sphere	cube	cone	other

22	Identifies various three dimensional figures (prisms, cubes, spheres, cones, cylinders, polyhedra, etc.)

Silly Cylindrical Civilians
Grade Level: Middle

MATERIALS: Cylindrical containers (oatmeal cannisters, cardboard juice containers, toilet paper tubes, etc.) flex straws, toothpaste caps, macaroni etc.

ORGANIZATION: Individually

PROCEDURE: In order for the students to understand the properties of a cylinder, they need to experience them in many sizes and lengths. This assignment can be done as homework or items can be assembled in class.

Begin by telling the students they are going to create cylinder people with the items they bring from home. The people must have arms, legs, eyes, noses, teeth and body. Using glue, scotch tape, etc. put the items together to make a SILLY CYLINDRICAL CIVILIAN.

You might also choose to have different groups of students use different shapes to create a person. Examples might include the Great Cone Man, Spheroidman, Mr. & Mrs. Polyhedron, or Sir Cube and the Cubettes.

As an extension, have your students write stories about them.

22	Identifies various three dimensional figures (prisms, cubes, spheres, cones, cylinders, polyhedra, etc.)

What Has Three...?
Grade Level: Middle

MATERIALS: Sets of pictures showing geometric figures (or a large set of actual geometric shapes), a bag or box for the pictures

ORGANIZATION: Whole class, small groups, or teams of two

PROCEDURE: Begin this lesson by modeling with the whole class. The teacher selects a picture of a geometric figure from a box or bag and does not show it to the class. The teacher then tells the class a riddle about the figure. An example might be, "This figure has six faces, each corner is a right angle, and each side is the same length. What is it?" The students try to guess what the figure is (a cube).

After the students understand the procedure, allow each student to make a selection from the bag of pictures. Make sure that they do not let other students see their picture. Give the students time to examine their picture privately and to write a riddle about it.

The students may read their riddles to the whole class, to a small group, or just one other student. The listeners then try to guess the geometric figure that the riddle is about. The riddles can be shared with other classes.

You might choose to use small actual shapes in the bag rather than pictures of the shapes. This provides students with a more concrete method of analyzing their shapes.

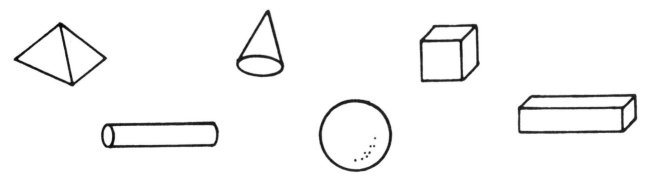

22	Identifies various three dimensional figures (prisms, cubes, spheres, cones, cylinders, polyhedra, etc.)

Clay Play
Grade Level: Middle

MATERIALS: A small amount of modeling clay for each student, art prints of landscapes, sheets of construction paper or tagboard

ORGANIZATION: Individually or in groups of three or four students

PROCEDURE: In this activity, students will be identifying three dimensional figures in a landscape drawing and will then try to re-create the scene in three dimensions with modeling clay.

Begin by displaying two or three landscape prints. It is nice to provide a variety, from mountainous scenes to rolling hills. Ask students to imagine that they are in the picture. Elicit responses as to what would be closest to them, what would be farthest away, what the various features in the print would look like from a top view, and how far apart the various features would be from foreground to background. This is a sophisticated discussion and will need much teacher direction.

Once students understand these concepts, tell them that they are going to re-create one of the paintings from a top view. The medium they will use to do this with is modeling clay.

If your class works well in cooperative groups, you might divide them at this point. You might also choose to allow students to do this project individually. As students work, you will need to circulate from group to group correcting students and questioning them as to the shapes which they are using in the creations.

The final product is secondary to the thinking process which students will go through to determine the relationships and ratio of each feature of the painting. You may find that this activity can be repeated several times during the year and students will continue to enjoy and improve their performance in the activity.

23	Identifies the properties of various three dimensional figures: length, width, height, side/face, base, vertical, edge, volume, and surface area

Six Faces Have I
Grade Level: Middle

MATERIALS: Boxes from home, construction paper, crayons, and marking pens

ORGANIZATION: Teams of two

PROCEDURE: Ask students to bring in empty boxes of various sizes and shapes (you may want to limit the size).

Have the students cover their boxes with paper to create a "jacket" of paper for the box. They may use a variety of methods to accomplish this and may need to use glue or tape in addition to the listed materials. On the box jackets, the students should mark the different geometric elements, such as edges, faces, etc. with a different color crayon or marker.

They should then trade their covered box with a partner and quiz one another to match the names of each property with its color. (i.e. red shows points, the faces are blue).

This is a basic lesson but provides a "self-created" model to help students remember or re-learn the concepts identified in the task analysis.

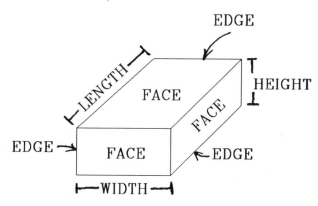

23	Identifies the properties of various three dimensional figures: length, width, height, side/face, base, vertical, edge, volume, and surface area

What You Got in the Bag, Lady?
Grade Level: Middle

MATERIALS: Bag or box of assorted junk such as old keys, pencils, coins, etc.

ORGANIZATION: Whole class activity

PROCEDURE: The teacher should prepare the junk bag before the lesson. The junk could be brought by the students or be any collection of items found in the classroom.

The teacher begins the lesson by drawing an item from the bag and showing it to the class. The class should state what geometric figure is suggested by the item and should be able to support their reasoning.

A sample student (or combined) response might be: "A pencil could suggest a circle because the eraser is round," or "a hexagon because it is 6 sided," or "a point because of the point you write with," or "a line segment because it starts with the point and ends with the eraser."

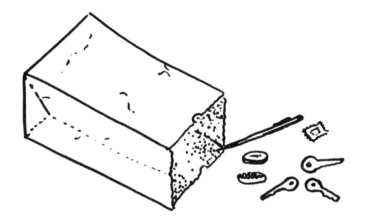

As students become more sophisticated in their answers, have them begin to identify such things as: the number of vertices in the junk item; the location of the height, width, and depth; or the estimated volume or surface area of the item. This portion of the lesson is well suited for cooperative groups working together to come up with as many possible discoveries about each object as they can.

23	Identifies the properties of various three dimensional figures: length, width, height, side/face, base, vertical, edge, volume, and surface area

Keeping Your Perspective
Grade Level: Middle

MATERIALS: Drawing paper, rulers, pencils, sample drawing showing perspective

ORGANIZATION: Individually

PROCEDURE: This lesson is an excursion away from math per se, but it is very helpful in showing the students how two dimensions become three through the use of perspective. You may want to integrate this lesson into an art period while students are working on geometry in math.

Begin by eliciting the properties of a one, two, and three dimensional figure from the students, i.e. one dimension has only length, two dimensions have only length and height, and three dimensions have length, height and depth.

Ask students if they know of a way to create depth on a two dimensional surface such as paper. Eventually you may have to introduce the terms "perspective," and "vanishing point."

Begin by drawing a square on the board and creating the vanishing points as shown. Let students experiment with this. You will find them fascinated with this process. For students or classes that understand the cube drawings, have them create a simple street scene.

For many of your students, this kinesthetic approach to actually creating a "third dimension" will help to cement the concept of dimensions into their thinking.

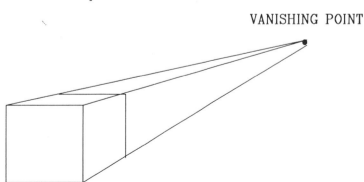

VANISHING POINT

23	Identifies the properties of various three dimensional figures: length, width, height, side/face, base, vertical, edge, volume, and surface area

Boxing Yourself In
Grade Level: Middle

MATERIALS: 1 cm or 1/4″ graph paper, tape, crayons, shoebox (or other appropriate objects), scissors

ORGANIZATION: Individually

PROCEDURE: In this activity students will be creating a three-dimensional figure and identifying its component parts.

Begin with a review of the three different elements of a three-dimensional figure: length, width, and height. Use the shoebox or other appropriate object as an example. Ask your students if they could put together a 3-D figure if they were given the parts. Explain that a three-dimensional shape is actually a set of two-dimensional shapes "connected" to show "depth." Ask them to identify the various shapes they see in the shoebox: two squares (one on each end) and four rectangles (the box top should be taped on).

At this point, the lesson becomes a bit more complex. Hand out a few sheets of graph paper to each student, and have them draw a two dimensional shape (a square or rectangle work best). Cut out this shape. On this figure, have them identify the height in one color and the length in another color (as shown below -- dotted vs. broken lines). They should then cut out five more of these same shapes and attempt to tape them together to form a three dimensional shape.

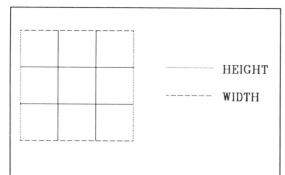

Most students will have drawn (cut out) a set of shapes that can't be made into a three dimensional figure. They will discover this soon enough. Ask them what they need to do to create the sides in the correct proportion (ie., cut a piece with the same length but different width for the top of a rectangle). Allowing students to make discoveries like this is valuable.

24	Sorts and categorizes figures and their elements into the appropriate dimension (one, two, or three)

ABC's of Geometry
Grade Level: Middle

MATERIALS: Paper, colored pens or markers, various ABC books for demonstration

ORGANIZATION: Individual or in small groups

PROCEDURE: Begin by telling students that they are going to create an ABC book of geometry terms similar to picture books on the ABC's that they have seen in the past.

Each letter and drawing can be done on a separate page, or you may have students do several letters per page. Each letter should be a geometry word and definition as well as a drawing of the item. If you are doing this activity with older children, you might do a cross-grade level lesson with fifth grade children and second grade children teaming to work together to make the book.

A sample set of terms of each letter is given below.

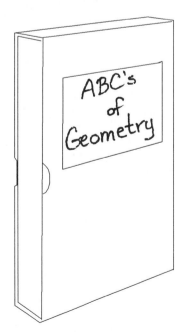

A – Arc; B – Bisect; C – Circle; D – Diagonal, Degree; E – Edge, Ellipse, Equilateral; F – Face, Figure; G – Geometry; H – Hexagon, Heptagon, Hypotenuse; I – Intersect, Isosceles; J – Join: L – Line, Line Segment; M – Midpoint, Moebus Strip; N – Nonagon; O – Obtuse, Oblique, Octagon; P – Parallel, Parallelogram, Pentagon; Q – Quadrant; R – Radius, Rectangles, Rhombus; S – Scalene, Secant; T – Tangent, Transversal, Trapezoid; U – Unequal; V – Vertex, Vertical,; W – Width; X – Four angles together, the intersection of two lines; Y – ; Z – Zenith, complementary angles, diagonals connected by two parallel lines

24	Sorts and categorizes figures and their elements into the appropriate dimension (one, two, or three)

A Monopoly on Figures
Grade Level: Middle

MATERIALS: Sets of pictures showing geometric figures

ORGANIZATION: Whole class or teams of two, three or four students

PROCEDURE: The students should be fairly familiar with geometric terms before playing this game. The pictures should be in a container so that the students cannot see them.

The teacher begins by selecting a picture of a geometric figure from a box or bag without letting the class see the figure. The teacher keeps the picture hidden while the class asks questions to guess which figure it is. For example, "Does it have four sides?" or "Is it one dimensional?" All questions should be answered yes or no.

When the class understands the procedure, they can play the game in small groups or teams of two, three, or four. They should take turns drawing pictures from the container and asking questions.

Actual geometric objects can be used instead of pictures.

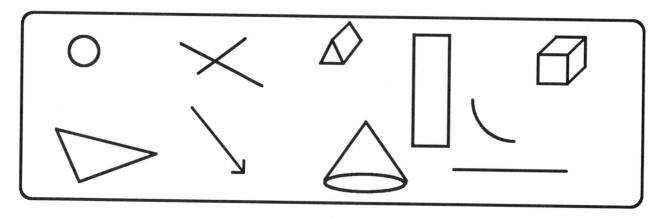

24	Sorts and categorizes figures and their elements into the appropriate dimension (one, two, or three)

Hando
Grade Level: Middle

MATERIALS: HANDO cards (see Appendix A), sets of pictures of geometric figures, beans or markers

ORGANIZATION: Whole class

PROCEDURE: Students should be familiar with geometric terms before playing this game. Have the students print geometric terms carefully and neatly in random order on their Hando sheets.

Hand out the beans or markers and begin by holding up a picture of a geometric figure or the object itself. The students should cover the word on their HANDO card which describes the displayed item as they would in regular bingo. The winner is the first student to have five in a row -- vertically, horizontally, or diagonally.

This game can be reversed with the figures drawn on the HANDO cards and the leader calling out the name of the figure.

H	A	N	D	O
		FREE SPACE		

| 24 | Sorts and categorizes figures and their elements into the appropriate dimension (one, two, or three) |

Tally Ho!
Grade Level: Middle

MATERIALS: Pencils and tally paper

ORGANIZATION: Teams of two or individually

PROCEDURE: This activity can be done in the classroom, on the playground, or as a homework assignment.

Hand out the tally paper and give directions. Students are to find as many geometric figures as they can in a given period of time; however, they must also classify them into one, two, or three dimensions. Winners are those who find the most figures, and/or those who have the greatest variety of figures.

One Dimensional		Two Dimensional		Three Dimensional	
Object	Location	Object	Location	Object	Location

25	Identifies elements and constructs representations of one dimensional elements and their symbols: point, line, line segment, ray, arc, and angles

What a Card
Grade Level: Upper

MATERIALS: 3 x 5 index cards, marker pens

ORGANIZATION: Whole class activity

PROCEDURE: Begin the lesson by giving each student 5 to 8 index cards (depending upon the number of vocabulary words you wish to review). Name a vocabulary word from the list below and ask each student to draw the symbol for the word. Have students hold up their cards for your inspection, correcting those which are are incorrect. Continue this process for each review word. An alternative method is to give a definition for the word rather than saying the word itself. In essence, this is the review part of the lesson.

The next step is to have students spread the assortment of cards on the desk in front of them. The teacher then points out or describes different items in the room while the students hold up the card which describes the specified item. Some samples might include the light emanating from a bulb, the corner of the chalkboard, a dot on a poster, etc.

Once students are comfortable with this, have each student generate a list of items in the room (or outside) which they might ask the entire class to identify. You might include the rule that they must have one item for each vocabulary card.

Students can either share their lists with the whole class or they might form cooperative groups to go over one another's lists.

Point
Line
Line segment
Ray
Half–line
Parallel
Perpendicular
Intersect

25	Identifies elements and constructs representations of one dimensional elements and their symbols: point, line, line segment, ray, arc, and angles

Instructodraw
Grade Level: Upper

MATERIALS: A variety of simple pictures, pencils, and paper

ORGANIZATION: Groups of two students

PROCEDURE: This lesson provides a reinforcement of communication skills in addition to a fun lesson on drawing various lines and shapes.

Begin by dividing students into teams of two and give one picture to each class member. Students should not share their pictures with one another. Pictures should be simple, without people, faces, or complex shapes. Certain cartoons and advertisements will work well.

One student in each twosome will attempt to draw the picture by listening to oral commands of his partner who is looking at the picture A sample explanation might sound like...

"Draw an oval shape near the bottom of the page but leave the top of the oval open."

Once the student has completed the drawing partners should compare the results. The students then switch roles and attempt to draw a second picture.

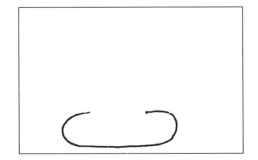

25	Identifies elements and constructs representations of one dimensional elements and their symbols: point, line, line segment, ray, arc, and angles

Rainbow Geography
Grade Level: Upper

MATERIALS: Map, crayons, protractor, pencil

ORGANIZATION: Groups of two or three students

PROCEDURE: This activity uses highways and roads on a map to help students identify the vertex of an angle, and the types of angles, acute or obtuse,

Begin by giving out a U.S. mileage and driving time map which can be found in most road map books. Any section of the United States may be enlarged or used as is. The teacher will tell the students what city to use as a vertex, then have them locate and then color specific items in different colors as follows:

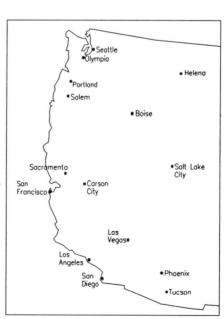

NAME A SPECIFIC CITY AS A VERTEX AND FIND;
      An acute angle--------color it red
      An obtuse angle--------color it blue
WITH ANOTHER CITY AS A VERTEX FIND;
      Two adjacent angles--------color them yellow
FIND A RIGHT TRIANGLE AND COLOR THE HYPOTENUSE BLACK
    AND THE BASE BROWN.
LOCATE A RECTANGLE WITH A DIAGONAL LINE
    Color the rectangle orange, the diagonal line purple.
FIND COMPLEMENTARY ANGLES
    Color them pink and green.
FIND TWO CONGRUENT ANGLES
    Color them grey
FIND AN ISOSOCLES TRIANGLE------color it blue-green.

25	Identifies elements and constructs representations of one dimensional elements and their symbols: point, line, line segment, ray, arc, and angles

Take Time for This Line
Grade Level: Upper

MATERIALS: A long piece of string, 3 by 5 cards, and a measuring tape

ORGANIZATION: A whole class activity

PROCEDURE: Time lines are used by most teachers during their social studies or history lessons. In this activity, you can integrate the concept of an existing time line in your classroom to the study of geometry.

Begin with a discussion of time, past and future, and elicit from your students that time is a constant progression or line of events. You might have students discuss the time lines of their lives, the time lines of a century, or the time line since the beginning of time. The A.D./B.C. transition can bog down a discussion on time lines because many students have difficulty uderstanding the relationship of A.D. to B.C.. We suggest you use a time line from the year 1 A.D. to the year 2000.

At this point, suspend a string across the length of the room, measure it, and have students figure out the value of one inch (cm), one foot, and one yard (meter) segments in terms of years. The teacher should then model the process by writing two or three historical events on an index card and having students help place the cards on the proper time line location.

Have each student write an event on a card and place it on the time line.

At this point, begin discussing the concept of point (the event on the time line), line segments (the entire time line from 0 – 2000 as well as the time between each event), rays (the time line extended back into B.C. and the beginning of the earth), and lines (the time line extended forever into the past and the future).

This lesson is very sophisticated but provides a graphic display of the concept of one dimensional geometry as it relates to the fifth dimension of TIME.

26	Identifies and constructs representations of one dimensional elements: coordinates/axis points on a graph or grid

A Burlap Grid
Grade Level: Upper

MATERIALS: Small pieces of burlap for each student, needles, thick string or yarn

ORGANIZATION: Individually and in teams of two

PROCEDURE: In this activity, students will use the warp and weft of a piece of burlap to represent the y and x axes of a grid. In order to do this activity effectively, students should feel relatively confident in working with x and y coordinates.

Begin by giving each student a 12" by 12" piece of burlap. Ask them to notice the way in which it is made -- interlocking vertical and horizontal threads. Elicit the response that this structure is very similar to a sheet of graph paper.

Have the students select a vertical strand near the left edge of the burlap and a horizontal strand near the bottom edge of the burlap. You may wish to have them color these strands with chalk or a crayon. These strands become the x and y axes.

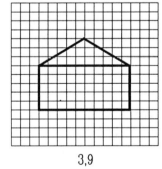

There are numerous ways to proceed at this point. You might have the students plan a design of some type. Very simple designs might be various geometric shapes while more complex designs might include flowers, houses, or simple landscapes. Students can stitch these patterns and then write the pattern so other students can duplicate their stitchery. Another approach would be to hand out graph paper and let students pre-draw the stitchery and write a plan to give to a partner who will try to reproduce the graphed design. A third possiblility is to use dot to dot designs which you can find in books at the local variety store. Students can try to duplicate the dot to dot pattern by writing the coordinate points. Several other extensions will occur as you watch your class begin the process. The important feature of this lesson is for students to write a coordinate plan of their drawing. A very elementary sample is shown.

3,9
13,9
13,4
3,4
3,9
8,12
13,9

26	Identifies and constructs representations of one dimensional elements: coordinates/axis points on a graph or grid

Angling In On the Treasure
Grade Level: Middle/Upper

MATERIALS: 1" by 1" (1 centimeter) grid paper, protractors, rulers

ORGANIZATION: Groups of two, three, or four

PROCEDURE: In this activity, students will use rulers and protractors to attempt to find a buried treasure on a grid.

Begin by drawing a ten by ten grid on the chalkboard and label the vertical axis A, B, C... and the horizontal axis 1, 2, 3... On a small sheet of paper, write the location of the treasure (F6 is the location we will use in the explanation). Begin the treasure hunt at the lower, left hand corner of the grid (A1) and tell student that the first clue is: draw a six inch line at a 30⁰ angle to the horizontal axis. Using your protractor and ruler, draw this line for the class and label the line segment ZY (using the alphabet backwards will alleviate some confusion.

The second clue is to draw a six inch line from point Y at a 150⁰ angle. Complete this task for the class and this line will be labeled YX. The third clue is to draw a five inch line from point X at a 20⁰ angle. If you have drawn correctly, you have found the treasure.

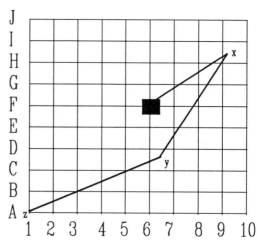

Divide students into groups and pass out rulers and protractors. Give them an opportunity to solve the treasure hunt listed below. Once a group has solved the treasure hunt, let them create their own hunt for the rest of the class to try. They will need to write very clear and structured directions.

26	Identifies and constructs representations of one dimensional elements: coordinates/axis points on a graph or grid

The Plot Thickens
Grade Level: Upper

MATERIALS: Graph paper, four decks of cards (remove face cards or give them different number equivalents), crayons or markers

ORGANIZATION: Pairs of students

PROCEDURE: In this activity students will be plotting points on a grid and, by asking a series of yes and no questions, will be identifying an unknown shape.

Begin by having students draw a line down the left edge of their grids another line across lower edge (as shown) label the vertical the Y axis and the horizontal line the X axis. The pairs of numbers they will be getting should be recorded with the X coordinate first (X, Y).

Model on the board the following procedure for your students. Draw two playing cards and plot them. Draw three more pairs and plot them also. Now, connect the points so that you have a quadrilateral.

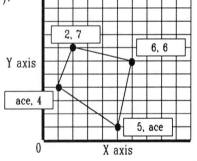

Explain that each student's task will be to graph four coordinate points to create a quadrilateral. They will keep their creation to themselves. The next step is for each student to try to re-create the partner's shape by asking yes or no questions to determine the coordinates.

For the example shown, sample questions and answers might be: "1,1" -- "outside;" "5,1" -- "outside;" "5,2" -- "inside;" etc. The guessing student will need to keep track of the location of guesses on graph paper and eventually will begin to see a pattern emerging as to the partner's hidden shape. The student using the fewest guesses is the "winner."

26	Identifies and constructs representations of one dimensional elements: coordinates/axis points on a graph or grid

Fire When Gridley, Ready?
Grade Level: Upper

MATERIALS: Several sheets of 1" (1 centimeter) graph paper

ORGANIZATION: Teams of two students

PROCEDURE: This is a sophisticated lesson in which students will need to communicate effectively with one another to complete the task.

Begin by dividing the class into teams of two. Each student should begin with one sheet of 1" graph paper and should number the horizontal and vertical axes. Student 1 begins by drawing a geometric shape on the graph paper. The shape should with the actual lines on the graph as much as possible (see examples). The student should keep this drawing hidden from the partner.

Begin the game with student two asking questions about which grid squares to color in. A variety of approaches can be used at this point. For example, you might limit the number of guesses students can make (12 might be a reasonable amount), you might do a "hangman" type of concept, or you might have both students do the "secret" drawing and then alternate guesses. The first student who identifies the actual shape is the winner.

Regardless of the format, proceed in the same manner. A sample game might be played as pictured:

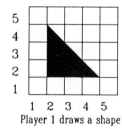

Player 1 draws a shape

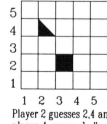
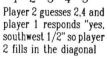

Player 2 guesses 3,2 and player 1 responds "yes" so player 2 fills in the square

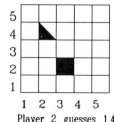

Player 2 guesses 2,4 and player 1 responds "yes, southwest 1/2" so player 2 fills in the diagonal

Player 2 guesses 1,4 and player 1 says, "no."

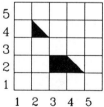

Player 2 guesses 4,2 and player 1 responds, "yes, southwest 1/2." Player 2 now has enough information to guess, "right triangle."

27	Identifies simple two dimensional elements and their symbols: angles (three letter notation), plane, and triangle

Make It Plane
Grade Level: Upper

MATERIALS: Crayons, paper, and pencil

ORGANIZATION: Groups of two

PROCEDURE: Explain to students that they will be going outside where they will look for angles, planes, and triangles. They are to choose a partner with whom they will work.

Take the students to a location where they will be able to see two sides of a building. Ask each team to work together to count the number of angles, triangles, and planes they see and to record the number. Next, have each group make a sketch of the building, leaving out details which do not contribute to the assignment.

Go back to the classroom and have each team redraw their sketch, this time, using a different color to show the triangles, the planes, and the angles. At this point, have the students recount the features and compare them to their original estimates.

Each of the triangles, angles, and planes may be labeled with letters so students may refer to them as they explain their drawing to the class or to another group.

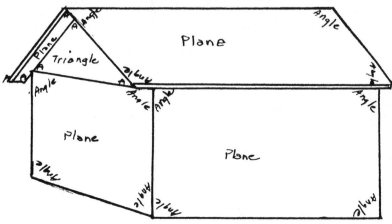

27	Identifies simple two dimensional elements and their symbols: angles (three letter notation), plane, and triangle

Shadows
Grade Level: Upper

MATERIALS: Any upright object on the playground such as a flagpole or tetherball pole, large protractor, directional compass (N,S,E,W), chalk or string

ORGANIZATION: Whole class activity

PROCEDURE: This activity must be done on a sunny day because students will be tracking and measuring the shadows created by the sun.

Begin this project as a whole class by going out to the playground first thing in the morning. Place a piece of tape or string along the shadow line of the school flagpole or tetherball pole. This will become the "base line" for measurement purposes.

Each hour, send a student or group of students out to mark and measure the number of degrees the shadow has moved from the base line. To do this, they will use the base of the flagpole as the vertex and will need a large (teacher demonstration size) protractor.

Create a chart on the chalkboard similar to that shown, and as the day progresses have students predict the next "shadow" angle.

At the end of the day have students discuss: the movement of the sun (actually the rotation of the earth); the relationship of the angles measured to a full circle of 360°; and the various uses of the sun's angles in terms of heating, growing periods, etc.

There are many extension activities possible including a unit of solar heating, navigation by the stars, rotation of the planets, and the seasonal changes.

Time	Degrees of Movement
8:00	
9:00	
10:00	
11:00	
12:00	
1:00	
2:00	
3:00	

27	Identifies simple two dimensional elements and their symbols: angles (three letter notation), plane, and triangle

This Letterman Has an Angle
Grade Level: Upper

MATERIALS: Uniform alphabet sheet (block letters work best)

ORGANIZATION: Cooperative groups of three or four

PROCEDURE: Explain to students that they will be exploring and identifying two dimensional elements such as triangles and squares. These elements can be found in our everyday environment, an example of this would be a simple alphabet sheet.

Divide the class into groups and explain that students will be playing a guessing game. Each team will be given a letter and will be giving and receiving clues in an attempt to guess the opposing team's letter from geometric clues. The teacher may wish to demonstrate with some of the sample letters from the bottom of this page.

The game starts with the first team giving a clue about their hidden letter. Team two should make a logical guess and then give their first clue to team one. Teams should have three to four clues per letter. For example, clues for letter "A" may include these statements: 1) "this letter has at least one isosceles triangle;" 2) "this letter does not have right angles;" 3) "this letter has only one triangle." Teams may wish to plan clues for several letters before the start of the game.

Game points suggestions: first guess, if correct, 5 points; second guess, three points; third guess, one point. Other sample letter clues: Letter "H" 1) this letter has no triangles, 2) this letter has at least one right angle, 3) this letter as at least three angles. Letter "B" 1) this letter has no squares, 2) this letter has at least two right angles, 3) this letter has four right angles.

27	Identifies simple two dimensional elements and their symbols: angles (three letter notation), plane, and triangle

Flight Plans
Grade Level: Upper

MATERIALS: United States map, protractors, and pencils

ORGANIZATION: Teams of two students

PROCEDURE: Begin by telling the class that in aviation, direction is given by degrees. The number of degrees is determined by the angle formed between the airplane's path and an imaginary north/south line. The angle is measured in a clockwise direction with north being the twelve o'clock position.

Have students figure the correct angle (number of degrees) for their flight plan from city A to city B. Students will ask numerous questions about how to hold the protractor, where the "base" line is, etc. Let them work together in teams to discover solutions to these problems.

Some of the clues you may want to give might include: there are 360 degrees in a circle; a protractor only measures 180 degrees; and the straight edge of the protractor needs to point a certain direction. Let students discover the meaning of these clues as they solve the work on this problem.

Once students understand the concept, let them quiz one another about flight pattern angles. You might extend this lesson to have students figure scale miles between destinations.

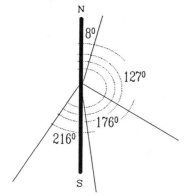

Fort Tundra -> Stenton
Fort Tundra -> Toledad
Fort Tundra -> Atlantis
Fort Tundra -> La Luce

28	Identifies various complex two dimensional elements and their properties: circles, polygons, trapezoids, parallelograms, and rhombuses

Keep Them in Stitches
Grade Level: Upper

MATERIALS: Thread, needles, and medium weight paper (which can be punched with a needle and will hold stitches), graph paper

ORGANIZATION: Individually

PROCEDURE: Students enjoy doing stitchery and needle work. This lesson incorporates their interest in these crafts to learn about relationships of complex geometric shapes.

A nice way to begin this project is to have students practice by making parabolic curves. This often done project is pictured below. This provides some background for the project at hand.

Begin by having students draw a simple design on graph paper. They should then select a colored piece of paper (the same size as the graph paper) and place them back to back. Have them punch holes with a needle through both sheets of paper at the appropriate locations. Give each class member a selection of various colors of thread. Students should be allowed to experiment with various shapes and not necessarily be confined to the symmetrical type of pattern shown.

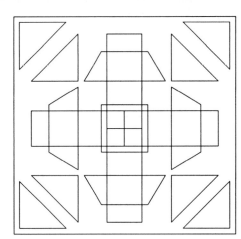

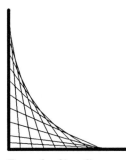

Parabolic Curve

Some students will want to do different shapes in different colors, while others will create random patterns with their shapes. Some will want to do several different projects as their design skills improve. Some students may find this easier if they glue colored paper to a piece of cardboard before beginning.

28	Identifies various complex two dimensional elements and their properties: circles, polygons, trapezoids, parallelograms, and rhombuses

Shapely Shapes
Grade Level: Upper

MATERIALS: Various construction paper shapes (triangles, rectangles, parallelograms, trapezoids, and rhombuses)

ORGANIZATION: Teams of two students

PROCEDURE: This activity is similar to working with tangrams except that pieces need not adjoin, they can lay on top of or overlap one another. A set of sample shapes and sizes are shown below. The purpose of the lesson is to give students an opportunity to discover relationships between various shapes.

Begin the explanation of this lesson by making a differentiation between complex shapes and simple shapes. In this case, simple shapes include the shapes shown in thin lines below, complex shapes are the multi-sided shapes which the simple shapes need to fill.

Hand out a duplicated set of the figures shown below and let students cut out shapes. Give them time to devise solutions. Once students understand the concept, let them try to create their own complex shapes and then exchange these shapes between teams or try to stump their teammate.

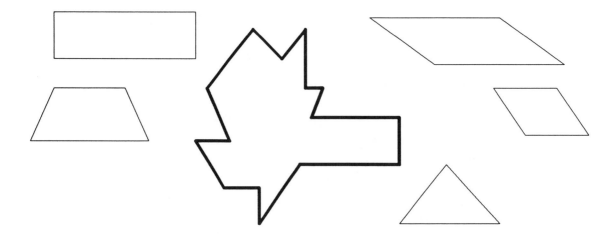

28	Identifies various complex two dimensional elements and their properties: circles, polygons, trapezoids, parallelograms, and rhombuses

I See the Light
Grade Level: Upper

MATERIALS: Four to six flashlights, movie screen, and a classroom that can be darkened

ORGANIZATION: A whole class activity with students taking turns to show various properties

PROCEDURE: This lesson is limited only by the creativity of your instructions and that of the students using the flashlights. Use this activity as a lesson to reinforce previous learning and investigate new information. It takes team work and patience.

Pass out the six flashlights to selected students. Darken the room and go through the following list of instructions:

FLASHLIGHT #1:	Make a point on the screen
FLASHLIGHT #1 & 2:	Make an imaginary line segment by showing its two endpoints
FLASHLIGHT #1, 2 & 3:	Make an imaginary right triangle by showing its three vertexes
FIASHLIGHT #1, 2, 3, & 4:	Make an imaginary parallelogram whose length is taller than its width

Ask students for ideas and directions to give to the demonstration group.

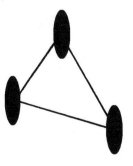

| 28 | Identifies various complex two dimensional elements and their properties: circles, polygons, trapezoids, parallelograms, and rhombuses |

Hear, See, and Speak No Evil
Grade Level: Upper

MATERIALS: No special materials are needed

ORGANIZATION: A whole class activity

PROCEDURE: In this activity students will be using their knowledge of two dimensional figures to create shapes. You may want to begin with a review of the elements of all the two dimensional figures.

Tell the students that you are going to designate four people who will be vertices. Then ask the whole class to work in cooperation to form a given geometric figure. You may tell them the name of the shape or simply give them the definition (a circle, or a figure in which all the points along the edge are equidistant from the center). It will be up to students to form the shape using everyone in the class. The one catch is that they cannot speak.

After they have acquired some skill doing this, add the stipulation that they are not only to refrain from speaking, but are not allowed to gesture with their hands.

Other variations include: 1) not specifying which students are to act as the vertices, 2) giving them a time limit, or 3) breaking them into teams of 12 or more and having a friendly contest.

29	Identifies relationships among simple two dimensional figures by analyzing vertex; acute, obtuse, and right angles; complementary, supplementary, and adjacent angles; diagonals, parallel, and perpendicular lines

Take Control, Air Control
Grade Level: Upper

MATERIALS: Large construction paper, rulers, markers, scissors, and glue

ORGANIZATION: Cooperative groups of four

PROCEDURE: Begin by discussing the geometric elements in the design of an airport. The lines created by runways, buildings, service roads, etc. create numerous interesting angles and line relationships. In this activity, students will design an airport incorporating the items listed in the task analysis.

Give each group of students an assortment of colored construction paper, rulers, markers, scissors, and glue and tell them that they will cut the various colors of construction paper to simulate airport features (shown below). The students' models must incorporate an obtuse and acute angle; supplementary and complementary angles; adjacent angles; and diagonal, parallel and perpendicular lines.

Upon completion, each group should explain their airport to another group, pointing out the various criteria listed above.

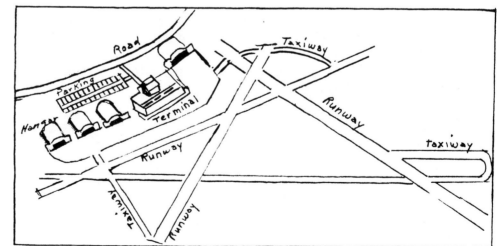

29	Identifies relationships among simple two dimensional figures by analyzing vertex; acute, obtuse, and right angles; complementary, supplementary, and adjacent angles; diagonals, parallel, and perpendicular lines

Timely Angles
Grade Level: Upper

MATERIALS: Classroom clock, or student-made clocks from paper plates

ORGANIZATION: Groups of two

PROCEDURE: Begin by reviewing complementary and supplementary angles with the class. Direct students to look at the clock and try to tell the times that would be shown if the hands formed 180° or 90° angles.

Divide students into groups and have them experiment with this question, with their ultimate goal being to create a table, chart, or some other representation which shows the various time formations. A sample approach and statement might be:

"1:10 and 2:20 are complementary angles (they form ninety degrees together)." "1:10 and 2:35 are supplementary angles."

As an extension, you might use the clock face for measuring other angles or for discussing the concept that a circle has 360°. Each five minute segment on the clock is 30°, each one minute mark is 6°, etc. You can also use the second hand to play a "name the angle" quiz game by comparing it to the hour or minute hand at any particular moment.

90 DEGREE ANGLE

180 DEGREE ANGLE

COMPLEMENTARY ANGLES

SUPPLEMENTARY ANGLES

29	Identifies relationships among simple two dimensional figures by analyzing vertex; acute, obtuse, and right angles; complementary, supplementary, and adjacent angles; diagonals, parallel, and perpendicular lines

ANGLE
Grade Level: Upper

MATERIALS: ANGLE card (similar to bingo -- see appendix C), spinner, teacher check card, beans

ORGANIZATION: Whole class activity

PROCEDURE: Begin by telling the class that this is a game similar to Bingo, called ANGLE. The teacher should then review the concept of complementary angles.

The game is played by the teacher listing various angle degrees (shown below) on the board in random order (i.e. 25°, 60°, 10°, etc.). Have the students place them in random order on their blank card with the word "FREE" in the center. Since there are only sixteen angles listed, the students must use some of them more than once. Be certain students understand that complementary angles are two angles which total 90 degrees.

A	N	G	L	E
		FREE SPACE		

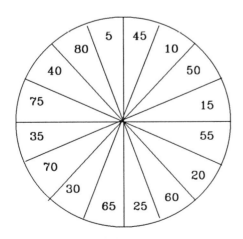

The teacher then spins the spinner and calls out the degree it points to. Students are to cover both that degree and its complementary angle degree (two beans for every spin) until a row, column or diagonal is covered with beans. At that time, the student will call out, "ANGLE." The teacher should have a card that shows the angles that have been called so the winner's card can be checked.

By placing different angle amounts on the spinner, this game can also be used for supplementary angles.

29	Identifies relationships among simple two dimensional figures by analyzing vertex; acute, obtuse, and right angles; complementary, supplementary, and adjacent angles; diagonals, parallel, and perpendicular lines

Color Wheel
Grade Level: Upper

MATERIALS: A regular hexagon, acrylic or tempera paint (or watercolors), brushes, rulers, pencils, jar lids

ORGANIZATION: Individually

PROCEDURE: Tell the students that their task is to combine the primary colors of adjacent angles to create secondary colors. This lesson is seemingly more of an art lesson than a math lesson, but once students have completed this exercise (emphasizing the term "adjacent angle") they will have a tangible and memorable frame of reference of the relationship of adjacent angles.

Begin by either drawing a regular hexagon or giving the students a dittoed hexagon. Tell students to draw diagonal lines to connect opposite vertices. They should have six angles when complete. Paint one angle with the color red. Leave the adjacent angle blank. Paint the next angle yellow. Leave its adjacent angle blank. Paint the next angle blue. Now three angles are colored with the primary colors and 3 are blank.

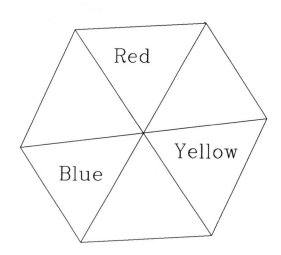

Direct students to combine two of the primary colors in a small jar lid to create the color of the adjacent angle which will be a secondary color.

Have students verbalize about the relationships of the primary to secondary colors and relate this to the fact that in this color wheel, these relationships are adjacent to one another.

30	Identifies relationships among complex two dimensional figures by ana- lyzing vertex; acute, obtuse, and right angles; diameter and radius; congruency, and interior angles

Getting the Angle On a Map
Grade Level: Upper

MATERIALS: City maps for each student (Appendix F has a San Francisco street map), protractors, rulers

ORGANIZATION: Groups of three or four students

PROCEDURE: Maps provide an excellent means of allowing students to label and identify angles. San Francisco provides a unique city street layout for this activity, but you may find other city maps (older cities tend to have odd shaped intersections and thoroughfares) appropriate as well.

Hand out maps to each group and have them label (A, B, C, etc.) each corner of several intersections. Students should then measure these intersections so they have a list of the number of degrees. At this point, students will write a set of directions for classmates to follow which will get them from one place in the city to another. In order to follow the directions, students will need to use a protractor and ruler.

Once each group has finished its puzzle, make copies and let classmates try to solve the puzzles.

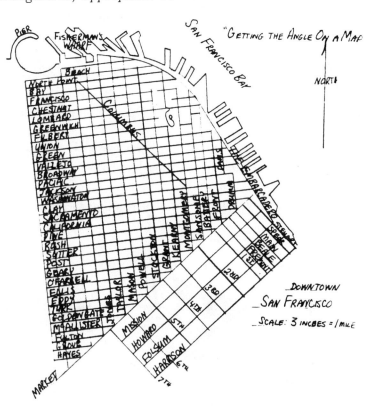

30	Identifies relationships among complex two dimensional figures by analyzing vertex; acute, obtuse, and right angles; diameter and radius; congruency, and interior angles

Letter Perfect
Grade Level: Middle/Upper

MATERIALS: Construction paper cut into 6" by 6" (15 cm) squares, scissors

ORGANIZATION: Individually or in teams of two

PROCEDURE: This is a lesson in congruency in which students will decide which direction to fold and then cut the paper squares so that they will form the capital letters of the alphabet. Congruent shapes are shapes which can be placed on top of one another and match exactly. You might point out that these letter shapes will also have symmetry.

Begin with a discussion of congruency with your students. Ask them if there are letters in the alphabet which are made up of congruent parts (see below). Tell students that they will need to make some decisions about folding paper and cutting out letters. Some letters can show congruency from left to right, and some from top to bottom. Some letters do not have congruent halves or parts.

Hand out a stack of construction paper squares and let students experiment with folding and cutting out the twenty-six letters of the alphabet.

For students who grasp the concept easily, see if they can fold paper both ways and cut out letters.

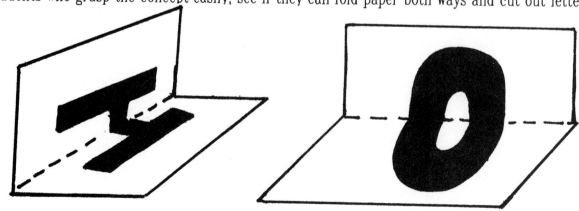

30	Identifies relationships among complex two dimensional figures by analyzing vertex; acute, obtuse, and right angles; diameter and radius; congruency, and interior angles

Amazing, to a Degree
Grade Level: Upper

MATERIALS: Paper, scissors, and rulers

ORGANIZATION: Individually

PROCEDURE: This lesson teaches the concept that there are 180 degrees in a triangle by having students cut out vertices of the angle and place them together to form a straight line.

Introduce the lesson by discussing the fact that a right angle has 90° and therefore two right angles (a straight line) would have 180°. Also mention that the students will find a relationship between a straight line and a triangle.

Give students a piece of paper and ask them to draw a triangle. They should draw it so it fills the paper and may be any shape (isosceles, scalene, etc.). Next, have them mark each angle as shown below, and then have students cut each corner away from the original triangle.

At this point, have students set up the triangle "corners" in a line as shown in figure 3. They will make the discovery that each triangle forms a straight line (contains 180 degrees).

You may wish to extend this lesson to do the same procedure with a square, rectangle, or parallelogram. You should elicit from students that 180° is a specific attribute of a triangle.

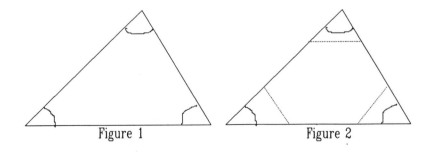
Figure 1   Figure 2

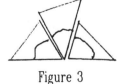
Figure 3

30	Identifies relationships among complex two dimensional figures by analyzing vertex; acute, obtuse, and right angles; diameter and radius; congruency, and interior angles

Try These Angles
Grade Level: Upper

MATERIALS: Protractors, scissors, construction paper

ORGANIZATION: Individually or in teams of two

PROCEDURE: Students may be able to say that a circle contains 360 degrees, but proof of this is another matter. This lesson provides a concrete example of this geometric reality.

Begin the lesson by drawing a triangle on the chalkboard and asking students if there is any way that triangles can be joined together to form a circular pattern. Let them ponder this question until someone comes to the conclusion that the only way triangles will fit together is if the triangles are isosceles or equilateral. Tell students that for this lesson they will be using only isosceles triangles.

Tell students that their task is to cut out a set of isosceles triangles which, when placed together, will form a circular pattern without gaps or overlaps. In order to successfully complete this task, students will need to realize that the center point of the triangle pattern will have to have 360° and therefore, the number of degrees in the vertex of each triangle will have to divide equally into 360.

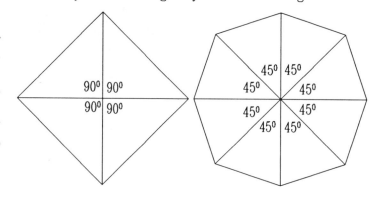

| 31 | Identifies and constructs various types of symmetry, transformations, and tesselations |

Where It Stops, Nobody Knows
Grade Level: Upper

MATERIALS: Tracing paper, chart paper, protractors, and pencils

ORGANIZATION: Individually or small groups

PROCEDURE: Rotational symmetry is a different aspect of symmetry which is interesting for students to understand. A figure with rotational symmetry matches itself exactly when rotated or turned less than a full turn.

The teacher should demonstrate by holding up a square and then rotating it 90, 180, and 270 degrees. Students can easily see that the shape is the same in each of these positions. This means that a square has the properties of rotational symmetry.

The next step is to show students that some shapes have rotational symmetry by being rotated more or less than 90°. A simple example would be an equilateral triangle or a five point star. In this lesson, students will be experimenting with creating rotationally symmetrical shapes by drawing a shape on a 6″ by 6″ piece of tagboard, tracing the space onto tracing paper and then using a brad (as if making a spinner) so the traced shape can be rotated. Students can then see if it aligns, and how many times it aligns with the original tagboard drawing.

Once they have created two or three shapes (spinners), have them compute the number of degrees the shape must be rotated to create symmetry. A sample shape is shown.

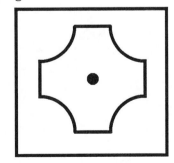

Tagboard with
shape drawn

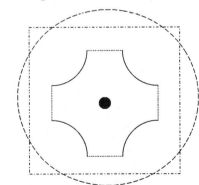

Tracing paper with
shape drawn

31	Identifies and constructs various types of symmetry, transformations, and tesselations

Kaleidoscope Art
Grade Level: Upper

MATERIALS: Crayons, design (see appendix H), kaleidoscope (optional)

ORGANIZATION: Individually

PROCEDURE: Begin by discussing what a kaleidoscope is and its function. Bring one in to demonstrate if possible, but it is not essential. Tell the students that what the kaleidoscope does is use symmetry with mirrors to project a variety of images. They are going to do the same thing by coloring the designs using symmetry.

This is not as easy a task as it may first appear, so have several extra designs for those who need to start over. There are certain rules:

1. Use only 3 colors (red, blue, yellow).
2. No color may be used next to the same color along the line segment.
3. The same color may only meet at the vertices.
4. The top must be symmetrical to the bottom, the right side to the left side

Display the finished pictures. Discuss the methods students used to create their designs and to keep them symmetrical.

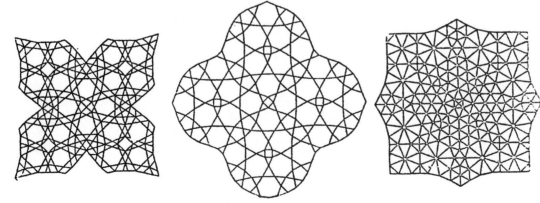

31	Identifies and constructs various types of symmetry, transformations, and tesselations

Mirror, Mirror
Grade Level: Upper

MATERIALS: Old telephone book (yellow pages), magazines, newspapers, scissors, and glue

ORGANIZATION: Individually or small groups

PROCEDURE: After students have gained a clear understanding of line and rotational symmetry, they can begin to consider symmetrical properties of trademarks.

Begin by reviewing the meaning of symmetry. Tell students that many trademarks of businesses are symmetrical. But their lines of symmetry may have:

> one line of symmetry
> two lines of symmetry
> three lines of symmetry
> point symmetry
> rotational symmetry.

Demonstrate what is meant by each kind of symmetry. Then have students look through materials and cut out one trademark that fits each kind of symmetry. Make a chart that shows the lines of symmetry for each trademark.

one line symmetry	two lines symmetry	three lines symmetry	point symmetry	rotational symmetry

| 31 | Identifies and constructs various types of symmetry, transformations, and tesselations |

Transformations
Grade Level: Upper

MATERIALS: Prepared codes, sheets of transformations (see appendix E)

ORGANIZATION: Small groups, teams of two, or whole class

PROCEDURE: The teacher should start this activity with the whole class. Pass out sheets of the code letters and the word to be decoded (see appendix E).

Work with the students to find the letters for the answer. They may need guidance at first to match the transformations with the figure for the letter. Discuss the ideas that the original figures for the letters and the transformations are really the same, just rotated or flipped.

When the students have guessed the answer to the example, have them put a word in code. They should trade with their partner and have him/her figure out their coded word. Later, have them write stories or riddles in code and have their partner figure it out.

They may want to make up other figures for letters and make up transformations for their own codes.

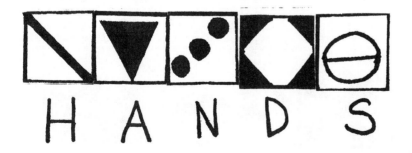

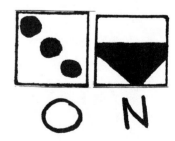

32	Identifies relationships among two dimensional figures to show ratio and proportion

These Models Have Scales
Grade Level: Upper

MATERIALS: Protractors, rulers, graph paper (optional), sample shapes, large sheets of paper

ORGANIZATION: Individually

PROCEDURE: Students will learn about ratio and scale in this activity by recreating shapes at one-half and twice the existing size.

Begin the lesson by drawing an example on the chalkboard of the shape shown. Measurements should be relatively accurate. Ask a student to come forward and measure the width and length of the various sides of the figure. Ask a second student to come forward and measure the various angles of the figure. Using the model on the board, have students attempt to re-create the shape at a 1:1 ratio at their desks (students may need graph paper).

Once this portion of the lesson has been completed, have students try to redraw the figure at one-half and then at twice the original size. Discuss the fact that the length of the sides should be doubled (or halved) but the angles will remain the same. If students have difficulty with this you may want them to work in pairs.

When they have completed their one-half and 2:1 scale models of the shape, hand out the shapes page and let them create a second shape to various scale sizes. For students who grasp the concept easily, you may wish to have them draw the shape to one-third or three-fourths of original size.

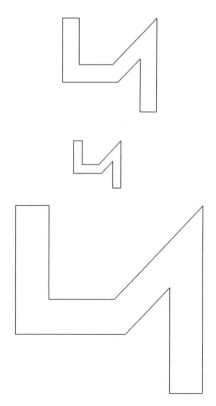

| 32 | Identifies relationships among two dimensional figures to show ratio and proportion |

As the Crow Flies
Grade Level: Upper

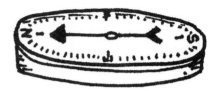

MATERIALS: Road maps, compasses, and rulers for each group

ORGANIZATION: Teams of two students

PROCEDURE: Your students may or may not be familiar with the expression, "as the crow flies," but this lesson will introduce them to a new use of the compass, a reinforcement of the term "radius," and a good lesson in measurement of miles.

Hand out a compass and a highway road map to each team of students and ask them to find an area on the map where several highways converge. It's best if they pick roads which are somewhat curved and irregular.

Next, have them locate the scale markings on the map (i.e. 1" = 10 miles, or 1 cm = 10 km). Have them set the compass to this marking and then place the compass point on the intersection of roads they selected and draw a circle. Double the length and draw another circle, add another length and make another circle. Do this until students have four or five concentric circles.

At this point they have hopefully encircled at least five to ten towns or destinations which they can compare the radius of distance "as the crow flies" to the actual distance in highway miles. Let them compute the differences in miles and share this information with another group.

The teacher should reinforce the use of the compass as a measuring device and the functional use of understanding the radius of a circle.

As an extension, you might include a discussion about the use of radius and circles in navigating on water or in the air. There are several computer games which reinforce this idea.

32	Identifies relationships among two dimensional figures to show ratio and proportion

The Incredible Bulging Planet
Grade Level: Upper

MATERIALS: Teacher constructed spaceship (use paper towel rolls, oatmeal cylinder, construction paper, etc.), similar materials brought by students

ORGANIZATION: Cooperative groups of four

PROCEDURE: This is a fun activity with many possible extensions for your class.

Tell students to imagine that they are on a space flight and have just landed on an unknown planet that they have named Blemo. As the spaceship touches down and the ladder lowers for the crew to disembark and explore, an unusual thing happens. The Blemoid atmosphere causes all crew members to grow to twice their normal size. As they look around them, they see that the entire environment is oversized and their spaceship has been dwarfed. This is the challenge of the lesson.

In order to escape from Blemo, the students will need to build a spacecraft exactly twice the size of the original (the teacher model). Give students time to plan and construct this model.

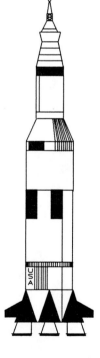

You may wish to assign a different ratio (proportion) to each group: twice, three times, or four times the original size; or one-fourth, one-half, or three-fourths the original size.

When students finish their models, have them display them for the class.

As an extension, have students write group developed stories about the great getaway from Blemo!

183

32	Identifies relationships among two dimensional figures to show ratio and proportion

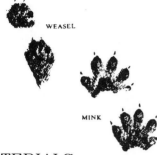

WEASEL

MINK

A Brief Paws
Grade Level: Upper

OTTER

MATERIALS: Graph paper (small and large), copies of paw prints

ORGANIZATION: Groups of two

2"

PROCEDURE: Explain to the students that within a family group of animals, sizes vary. From the footprint left by an animal, a person can identify which animal made the print as well as the size of the animal.

Give students a copy of the prints shown below (or have them find prints in books). To realize the actual size of the animal prints, have the students first trace the pictured prints onto small graph paper and then transfer the drawing onto large graph paper.

Depending upon the scale used in your source materials, you might have students make their own large graph paper. This in itself is excellent practice in identifying ratio and proportion.

Display the student drawings and then discuss various aspects of the prints, i.e. What would the relative weights of these animals be? What would the relative heights be? etc.

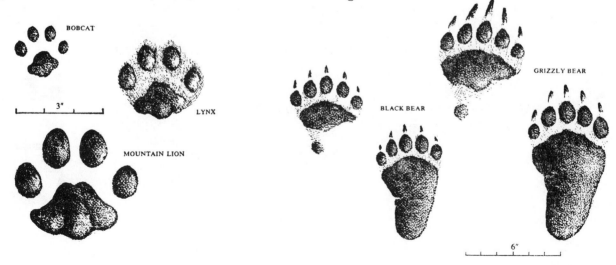

BOBCAT

LYNX

3"

MOUNTAIN LION

BLACK BEAR

GRIZZLY BEAR

6"

33	Uses various methods (compass, protractor, straight-edge) to construct two dimensional figures

Flower Power
Grade Level: Upper

MATERIALS: Protractors, compass, pencil, paper and crayons

ORGANIZATION: Individually

PROCEDURE: Begin by telling the class that a knowledge of geometry is helpful in making all sorts of designs in their art work, sewing, woodworking and decorating.

Many geometric designs are patterned after things that occur in nature, such as leaves of plants and trees, snowflakes, rock crystals and flowers.

The designs shown below are patterned after flowers pictured above the design. All four designs are geometric shapes. Ask students if they see the circles, the squares, the triangles, the right angles, the acute angles, the obtuse angles, the straight lines, and the curved lines.

Have the class bring in a flower and draw it in its geometric shape to create a design that could be used in an artistic creation.

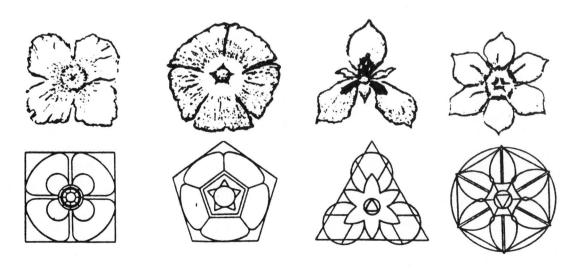

33	Uses various methods (compass, protractor, straight-edge) to construct two dimensional figures

Geometry -- From a Bird's Eye View
Grade Level: Upper

MATERIALS: No special materials are required

ORGANIZATION: A whole class activity

PROCEDURE: This is a lesson in group movement and cooperation as well as geometry. The rules for this activity are:
    1) Students in the group may not talk or give other directions.
    2) Speed is not important, cooperation and knowledge of geometry is the important feature.

Tell the students that they will be working together as a group to form a geometric shape which you, the teacher, will specify. You will draw a geometric figure on the chalkboard, they will be responsible for organizing themselves into the outline (perimeter) of the figure shown. Tell your students that you will be using points to represent where they should stand. They should think of the points as being the top, or bird's eye view of the figure.

The activity needs to be done in a space big enough for your students to move around freely. You might wish to take them outdoors.

After drawing a figure on the board, have your students form themselves into a representation of that shape. Ask them to identify the shape and discuss how they demonstrate its particular attributes.

33	Uses various methods (compass, protractor, straight-edge) to construct two dimensional figures

Picture Perfect
Grade Level: Upper

MATERIALS: String, chalk, tape measure for each group, a carpenter's "snap" chalk line (if possible), large concrete or asphalt play area

ORGANIZATION: Cooperative groups of three or four students

PROCEDURE: Students will learn about diagonals and triangulation in this activity as they attempt to draw a "perfect" polygon.

Begin this lesson with a discussion of why construction workers must build buildings which are perfectly "square" at the corners. Also discuss how they achieve this goal of "square" corners over long distances such as those in a house.

Tell students that in this exercise they are going to try to draw perfect polygons on the school playground. All they will have to use is chalk, a measuring tape, and string. You will probably want to do this activity two or three times with your class, so you might begin with squares and rectangles and then move to triangles, parallelograms, and multi-sided figures on other days.

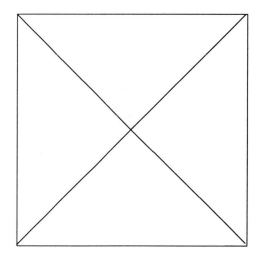

Draw a square on the board and ask how you could check that all sides and all angles are equal. Use of a ruler and protractor would enable you to do this on a small figure, but on a large figure the margin of error is much greater and the use of a protractor is inaccurate. Show the students that by measuring diagonally the square can be checked for perfection. Hand out materials along with a slip of paper which gives a specific size square or rectangle for each group to draw.

33	Uses various methods (compass, protractor, straight-edge) to construct two dimensional figures

Quilting Geometry
Grade Level: Upper

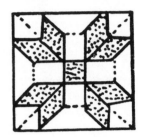

MATERIALS: Quilts or afghans from home, quilting books, paper, rulers, pencils, crayons

ORGANIZATION: Individually

PROCEDURE: Create a display of the quilts or afghans by hanging them around the room so that they can easily be seen. It may be interesting to have students relate the history of the quilt since many are considered family heirlooms.

Next, ask the students to study the design of the quilt. Their task will be to copy on paper the geometric figures used to create the pattern in the quilt. They should first trace the figures in pencil and then color them in using the crayons.

It is important that both the positive and negative backgrounds/spaces be studied to fully understand and appreciate the design of the pattern, some quilts can be very elaborate and deceptive. They should also note the repetitions involved.

After students have drawn their designs, they should create a chart listing the various geometric figures involved.

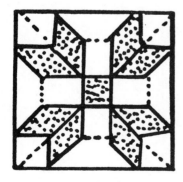

34	Identifies various three dimensional figures and their properties: sphere, cone, cylinder, polyhedra (cube, pyramid, and prisms)

3–D Mail Orders
Grade Level: Upper

MATERIALS: Catalogue, scissors, glue, paper for student charts

ORGANIZATION: Individually

PROCEDURE: This task requires each student to make a chart with various three dimensional shapes represented by items from a catalogue. Each student will display the shapes on a chart. The chart should be labeled sphere, cone, cylinder, cube, pyramid, prism, and rectangle. Under each heading, the student should find as many items that have that particular shape as he can, cut them out, and paste in the appropriate space.

sphere	cone	cylinder	cube	pyramid	prism

34	Identifies various three dimensional figures and their properties: sphere, cone, cylinder, polyhedra (cube, pyramid, and prisms)

Moving Geometry
Grade Level: Upper

MATERIALS: Construction paper, scissors, glue

ORGANIZATION: Individually or in groups of two

PROCEDURE: Begin this lesson by taking your class to the parking lot and let them take a look at a car or truck. Discuss the various geometric shapes which create the body of the vehicle. Have them analyze the lights, the grill, ornamentation, and all aspects of the automobile. You might even have them do a rough sketch of the vehicle.

When the children get back to class, tell them that they are going to try to do a side view and a front view of the vehicle by cutting out the shapes from construction paper and gluing them to a large piece of paper. Explain to students that they are making a two dimensional model of a three dimensional shape (the vehicle).

The students must be able to identify each shape they cut as well as its three dimensional counterpart.

As an extension, you might have them experiment with trying to imagine what the top view of the vehicle might be and try to recreate this view with various geometric shapes.

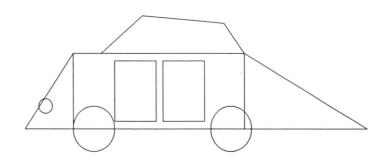
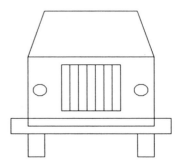

34	Identifies various three dimensional figures and their properties: sphere, cone, cylinder, polyhedra (cube, pyramid, and prisms)

Animals by Committee
Grade Level: Upper

MATERIALS: Various three dimensional shapes (boxes, cans, tubes, etc.)

ORGANIZATION: Cooperative groups of four

PROCEDURE: In this activity, students will work together to create a new animal and will then construct it out of three dimensional shapes.

Begin by dividing the class into cooperative groups of four. Tell students that their assignment is to create an animal for the new classroom zoo. You may wish to find some library books (primary) which describe such bizarre animals (Dr. Seuss is a good resource).

Give students time to collect (bring from home) various paraphernalia. Groups may want to divide the responsibilities and have each student in the group be responsible for designing one of the various parts of the body -- i.e. head, tail, legs, or body.

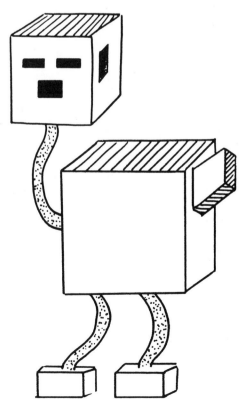

Once groups have finished their creations, have them name the animals and create a classroom zoo. Some sample names might include a Polyhedronicus, Rectangular Prismatorus, Cubazoid, etc.

As an extension, you may want to have the students write a story about their creations.

34	Identifies various three dimensional figures and their properties: sphere, cone, cylinder, polyhedra (cube, pyramid, and prisms)

Faces, Names, and Shapes
Grade Level: Upper

MATERIALS: 1 tetrahedron die, games sheet (Appendix G), one marker per player

ORGANIZATION: Groups of two or three

PROCEDURE: Begin by explaining the rules of the game. Hand out a game sheet to each group and describe how to play:

1. The object of the game is to land on the last shape on the game board.
2. The player whose last name contains the most letters plays first.
3. Take turns rolling the die. Move your marker the number of spaces shown on the die. If you can name the figure, the number of vertices, and the number of faces of the item in the box that you land on, you may keep your marker there. If you cannot name these features, you must return your marker to its previous position.
4. The first player to land exactly on the last shape is the winner.

Rectangular prism
6 faces
8 vertices

Hexagonal pyramid
7 faces
7 vertices

Oblique cylinder
3 faces

Rectangular pyramid
5 faces
5 vertices

Hexagonal prism
8 faces
12 vertices

Cylinder
3 faces

Square pyramid
5 faces
5 vertices

Triangular prism
5 faces
6 vertices

Cone
2 faces
1 vertex

Square prism or cube
6 faces
8 vertices
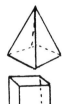

Triangular pyramid
4 faces
4 vertices
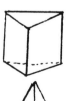
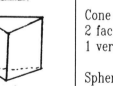

Sphere

35	Identifies relationships among various three dimensional elements: ratio and proportion, congruency, symmetry, base, edge, vertices, and side/face

Solve My Riddle
Grade Level: Upper

MATERIALS: No special materials are necessary although it is helpful to have sample models of various polyhedrons, a copy of the story The Riddle of the Sphinx

ORGANIZATION: Individually or in teams of two or three students

PROCEDURE: In this activity, students will practice using the number of faces, vertices, bases, and edges in given polyhedrons to create riddles for classmates. An excellent motivational technique is to read the Riddle of the Sphinx story to the children in preparation for this lesson.

Give each student a model or a picture of a polyhedron. These should include all types of prism shapes, pyramid shapes, and circular shapes. Tell the students to keep their shape secret from their classmates.

Have the students secretly study their polyhedron and try to identify the number of vertices, faces, edges, equivalent faces, the size of the angles, and all other information which you have discussed as a class.

Have students write this information down and then ask them to imagine the story of the Riddle of the Sphinx. Tell students to construct a riddle based upon the information they have compiled about their polyhedron. Clues must be accurate, but since it is a riddle, they must not be too obvious. Therefore, children will want to select clues carefully.

Once students have written their riddles, have them share them with the class. A sample riddle is given.

"I have four faces, and each face has the degrees of half a circle. All of my angles are acute. What am I?"

ANSWER: A triangular pyramid

35	Identifies relationships among various three dimensional elements: ratio and proportion, congruency, symmetry, base, edge, vertices, and side/face

No Distortion in Proportion
Grade Level: Upper

MATERIALS: Sets of cubes (student made cubes or sugar cubes)

ORGANIZATION: Groups of three or four

PROCEDURE: There are two approaches you can take in teaching this lesson. You can have students stack blocks to show the ratios and proportions given, or you might have students draw sets of cubes rather than using actual cubes. Regardless of your approach, the lesson format is the same.

Begin by drawing (on overhead) or holding up a cube. Ask students to describe the length, width, and height of this cube. Write the ratio as 1:1:1 and explain this structure to students.

Next, write several different ratios on the board involving 2 units (i.e. 2:1:1, 1:2:1, 1:1:2) and have students draw or construct these figures. Discuss with students the volume of each figure and the congruency. Students should eventually reach the conclusion that each of these figures is congruent (depending upon the view, top, front, side) and that the volume of each figure is the same.

Continue with this procedure and let students work with the figures represented by 2:2:1, 2:1:2, and 1:2:2.

As an extension, you might want students to create a table which describes the volume and congruency of each proportional figure.

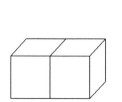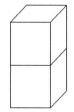

35	Identifies relationships among various three dimensional elements: ratio and proportion, congruency, symmetry, base, edge, vertices, and side/face

Planet Before You Begin Cutting
Grade Level: Upper

MATERIALS: Almanacs or encyclopedias for finding ratios between planet size, construction paper, tape, and scissors

ORGANIZATION: Cooperative groups of four students

PROCEDURE: This is a sophisticated lesson in which students will be constructing three dimensional models to represent ratio and proportion between the planets of the solar system.

Begin by asking students to do research to find the diameter of each planet in the solar system. Choose one planet on which to base your ratio (samples below show Earth and Pluto). This portion of the activity may prove to be difficult, therefore the completed information is provided below.

Once students understand the concept of ratio, assign each cooperative group a different geometrical shape, e.g. cube, rectangular prism, square pyramid, triangular pyramid, etc. -- (a sphere is too difficult to construct). Each group should then attempt to construct a model solar system of their shape. All planets should be done in the correct proportion.

When students have finished this assignment, you might have them write an explanation of the effects of living on planets of their shape. For example, what difficulties would one experience if the earth were cubical instead of spherical?

	Diameter	Ratio to earth	Ratio to Pluto (smallest planet)
Mercury	3000 mi.	.4:1	1.5:1
Venus	7000 mi.	.9:1	3.5:1
Earth	8000 mi.	1:1	4:1
Mars	4000 mi.	.5:1	2:1
Jupiter	89000 mi.	11:1	45:1
Saturn	75000 mi.	10:1	37:1
Uranus	31000 mi.	4:1	15:1
Neptune	30000 mi.	4:1	15:1
Pluto	2000 mi.	.25:1	1:1

35	Identifies relationships among various three dimensional elements: ratio and proportion, congruency, symmetry, base, edge, vertices, and side/face

Water Balloon Insight
Grade Level: Upper

MATERIALS: A funnel, measuring cup, water, and two balloons for each group

ORGANIZATION: Teams of two

PROCEDURE: This activity is best done outdoors or over buckets in case of accidents. Students will be comparing volume with the diameter of various spheres.

Begin by asking students to predict what will happen: if we fill a balloon with one cup of water, what will be its diameter; what will happen if we double the volume by using two cups of water, then three?

Now have the students begin the experiment by taking a balloon and fitting it over the nozzle of the funnel. One student should hold the funnel and balloon while the other pours. They do not necessarily need to tie off the balloon before measuring its diameter, they will also be less inclined to use the balloon in inappropriate ways. Once the balloon has been filled with the correct amount of water, have them rest it gently, still in hand, on a ruler. It is important that the balloon stay as round as possible. As it rests on the ruler, the partner should get two books and place them so that they just barely touch opposite sides of the orb (the books should be as parallel as possible). They should then remove the balloon and measure the distance between the books. This will give the students a roughly accurate measurement of the balloon's diameter.

Record results in the form of a table. Repeat this procedure with two and three cups of water and compare the results. A similar comparison can be made with the circumference by using the formula circumference = pi x diameter.

36	Uses various methods to construct three dimensional figures

Geomobile
Grade Level: Upper

MATERIALS: Straws, string, scissors, models of various polyhedrons, tissue paper (optional) coat hangers or thin dowels

ORGANIZATION: Groups of three or four

PROCEDURE: For some students, the best way to understand the properties of polyhedra is to actually make the shapes themselves. In this activity, students will use straws and strings to make various shapes and will then string the shapes together in a mobile.

Divide the students into teams of three or four and hand out materials. You will probably want to model the making of one shape for the students (if you prepare a few models in advance, students may have more fun trying to figure out how to make them without any extra instruction). Remind students that they are making three dimensional shapes.

Cubes and rectangular prisms are the most common shapes and once groups have completed these, let them experiment with pyramid shapes or perhaps a cone or cylinder. You may also want to have students cover the straw shapes with tissue paper (although this can be messy and time consuming). Once the group has completed several shapes have them construct a mobile similar to that shown.

| 36 | Uses various methods to construct three dimensional figures |

Personal Space
Grade Level: Upper

MATERIALS: Butcher paper, meter sticks, rulers, pencils, markers, and pens

ORGANIZATION: Groups of two

PROCEDURE: This lesson is somewhat abstract in concept, but it provides an opportunity for students to think in geometric terms in an unusual frame of reference.

Begin by saying that each of us has a "personal space" that others must stay out of if we are to remain comfortable. When someone comes too close, they invade this personal space. This amount of space and the shape of our space differs with individuals. The students' task is to identify and measure the size and shape of their own "personal space."

To complete this task, students will be working in teams of two by asking each other questions such as:
Does it bother you when someone talks to you from directly in front of your face?
Do you prefer having someone talk to you from the side or from behind?
Do you like it when people look into your eyes when they talk to you?
Does it bother you when people touch your arm or your hand when they are talking to you?

Have the class generate five or ten more questions to ask one another.

Once students have answered the questions, have them draw three figures of themselves, a front view, a side view, and a top view. Have them draw the shape that they think is important that others not invade. From these three drawings, have students create a three-dimensional figure which shows their personal shape.

36	Uses various methods to construct three dimensional figures

Three-D-oramas
Grade Level: Upper

MATERIALS: Small box (shoe box) per student, construction paper, scissors, glue

ORGANIZATION: Individual or small groups

PROCEDURE: Explain to your students that they will be creating a polygon world in a shoe box diorama (the teacher may wish to discuss a diorama). Students need to decide which special polygon their group wishes to explore for this assignment. Choices may include; cubes, rectangular prisms, cones, cylinders, etc.

The students then may begin to plan the "scene" their diorama will explore. Examples may include a city scene, classroom scene, store scene. etc.

As an extension, have students write a story describing their scene.

36	Uses various methods to construct three dimensional figures

You Can See It, But Can You Draw It?
Grade Level: Upper

MATERIALS: Overhead projector, a variety of small items to make shadow projections.

ORGANIZATION: A whole-class activity

PROCEDURE: In this activity, students will have to use their imaginations to draw a shape created from the shadow of a common object. Some items which cast unusual shadows are keys, rings, eyeglasses, spoons, forks, various kitchen utensils, hand tools, etc.

Begin the lesson by holding up an item and asking students if they can draw it. You will get a variety of answers, but try to elicit the response that it is difficult to draw because it is three dimensional and it is hard to make three dimensions look realistic. Continue the discussion to see if students might be able to draw the "shadow" of the item; ask why this would be easier. The response is that a shadow is two dimensional and is much easier to draw.

Ask students if the shadow of the instrument would be the same regardless of its position. Obviously, the shadow changes as the position of the item changes. Tell students that their task will be to draw the shadow of various items from three different positions, a top view, a side view, and a front view. The catch is that the students must draw the shadow without seeing the object. You will need to eventually put the item on the overhead so students can see how accurate they have been.

Have students draw a front, side, and top view of a hammer as though it were a shadow. Give them time to work with this. If students have a very difficult time, you might display the hammer to refresh their memories of its construction. Once they have finished drawing, lay the hammer on the overhead in these three different positions. Discuss the accuracies and inaccuracies of your students' drawings. Continue this process with several other items, each time emphasizing the concept of two versus three dimensions.

37	Sorts, classifies, and constructs various figures into one, two, and three dimensions

As Easy As 1, 2, 3
Grade Level: Upper

MATERIALS: Straws or spaghetti, glue or tape

ORGANIZATION: Individually

PROCEDURE: Explain to your class that you will be exploring the attributes of the three dimensions and constructing a model. Ask them to explain to you the difference between each of the three dimensions; ask for examples which can be put on the board.

Next pass around the straws and ask which dimension this might represent. They will, hopefully, identify it as one dimension. Now, ask what they would need to make the straw into a two dimensional object. Pass out enough straws for them to construct a figure. Next, ask what they would need to make their two dimensional figure into a three dimensional figure. Give them enough material (straws and tape) to complete the task.

The object here is to see how each dimension is constructed from elements of the previous ones, how each one is more complex than the one before, and how each dimension fills up a different aspect of space.

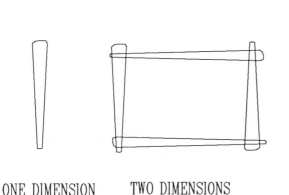 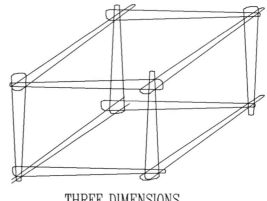

ONE DIMENSION   TWO DIMENSIONS   THREE DIMENSIONS

37	Sorts, classifies, and constructs various figures into one, two, and three dimensions

What You Got in the Bag?
Grade Level: Upper

MATERIALS: Grab bags of various items (one per group), butcher paper

ORGANIZATION: Groups of two to four

PROCEDURE: Explain to your class that in this exercise they will be identifying the items in each bag according to their relative attributes. The object will be to identify each item so accurately that others will be able to correctly tell what is in the bag without looking or being told.

Distribute a bag to each group and have them create a list of geometric attributes for each item. For example: item A -- a hollow rectangle (match box); item B -- a line segment (spaghetti noodle); item C -- partial plane inscribed with assorted intersecting line segments (business card). Have them write their list on a piece of butcher paper and tape it to the table next to the bag.

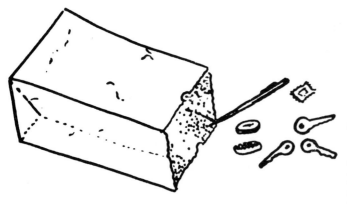

Now give students the opportunity to circulate among the tables and write in possible objects which would fit the descriptions of the items. The goal here is not necessarily to correctly identify the items, but rather to have the students realize the wide range of objects which can be described geometrically.

As an extension, discuss with the class how they might make this exercise more challenging. Some ideas might include putting more complex items in the bag such as paper clips or a Rubik's cube.

37	Sorts, classfies, and constructs various figures into one, two, and three dimensions

Have I Got a Deal for You!
Grade Level: Upper

MATERIALS: Boxes, cans, lids, toothpicks, and glue

ORGANIZATION: Individually or in groups

PROCEDURE: Explain to your students that they will be using geometric elements to construct a model of a car. They may create an imaginary car or a copy of a real one, but must use materials from home or the classroom (no pre-made parts) and the model must contain at least five elements from each of the three dimensions (you may need to vary the number of elements from each dimension).

Have them begin by drawing a rough draft of their automobile. Next, have them create a chart or list divided into three columns, one for each dimension. They should create two sub-columns in each of the dimension columns. Have them label the left-hand column as the actual part (axle) and the right-hand column as the geometric element (line segment). They must list all the parts used accordingly.

Once students are finished with their construction, have them create a display by taping the list of components onto the table next to their vehicle. Allow time for students to examine all the creations.

Next, explain that they will be pricing each car by the complexity of its component parts or by some other standard. For example: a circle is worth ten dollars, a triangular prism is worth 50 dollars, etc. They must then label and total the parts (this is how auto repair shops operate--for the most part). Some students may find that they can use their component list to reach their totals. Encourage them to use their own methods as long as parts aren't listed more than once. An interesting question is whether or not it is possible for a car to cost more if it is priced according to its components.

It might be fun to create a catalogue of various cars and their prices (a Polaroid camera comes in handy here). Another extension might be to see if they can do what the real car companies do-- construct a car aimed at a given consumer price range (economy, mid-range, and luxury).

37	Sorts, classfies, and constructs various figures into one, two, and three dimensions

Rooms of the Future?
Grade Level: Upper

MATERIALS: Markers, miscellaneous objects from home (match boxes, string, cans, pencils, candles, etc.)

ORGANIZATION: Groups of two or four

PROCEDURE: This exercise helps students to relate geometry to the world around them. They will begin to see how different geometric elements shape our surroundings. As they create 3-D models, they will identify the materials they use as either one, two, or three dimensional.

Explain to your class that they will be constructing something, and that they must bring all the materials from home (discourage doll house furniture or other prefabricated items). They are to build a model, more or less to scale, of a room, building, city block, or even the city itself. It may be real or imaginary (but within some realm of possibility, if they get too creative the geometric value might be sacrificed). They may build their own room, a school room, or what they would like to have when they are older.

After groups have been formed, have each student submit a rough draft of their construction along with a list of possible materials from which they might construct the furnishings. Have them collect their material for homework.

Next, have one of the students create a chart which is divided into three columns, one for each dimension. Further divide each column into two sub columns; one for items from home and the other for its geometric equivalent. Before beginning construction, have the students list all their materials in the appropriate column and identify its corresponding geometric element (eg., spaghetti = line segment/one-dimension).

After each group has finished construction, have students share their creations with classmates. Students should focus their sharing on the geometric shapes used to create their models.

H	A	N	D	O
		FREE SPACE		

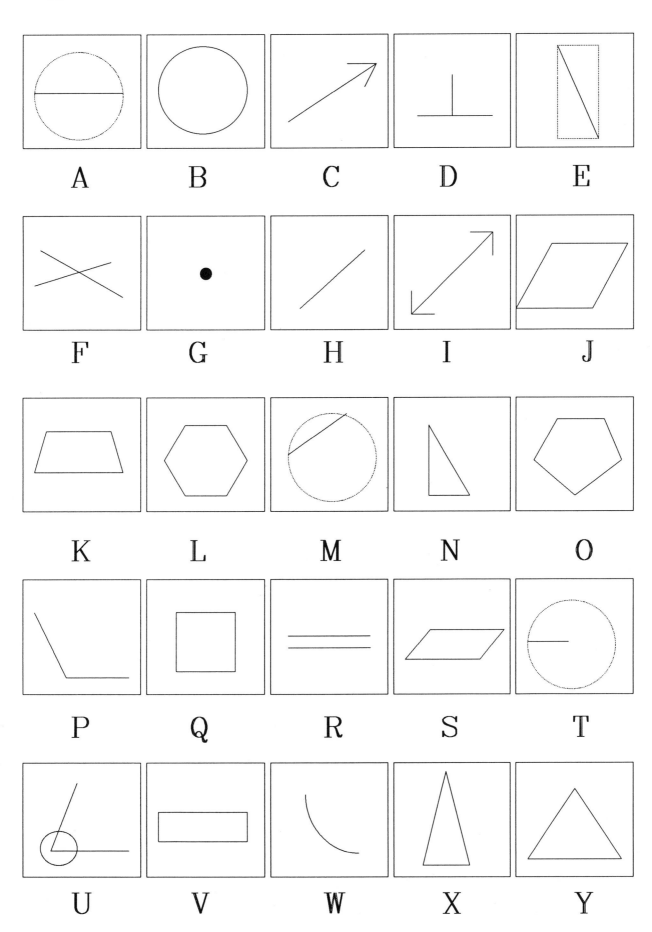

A	N	G	L	E
		FREE SPACE		

Making a Spinner

Many activities in this book require the use of spinners. Students can make their own spinners very easily and inexpensively by following the directions given here.

Materials required are: cardboard, scissors, rulers, pencils, paper clips, paper punch, and tape.

1. Cut out a cardboard (tagboard) pointer and punch a hole in the center.

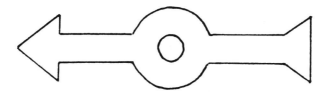

2. Cut a scrap of cardboard as a paper washer and punch a hole in the center.

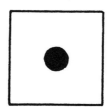

3. Cut out a four inch square of cardboard and divide it into quarters as shown. Make light pencil lines.

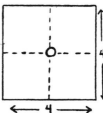

4. At the center of the square, make a small hole with your paper clip.

5. Using a compass, make a circle on the four inch square and draw the design you wish to use. You may wish to color your spinner at this point.

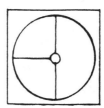

6. Bend the center loop of the paper clip up at a 90 degree angle to the outer loop.

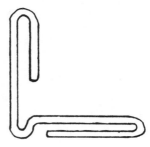

7. Tape the outer loop of the paper clip to the bottom of the four inch square to hold it in place.

UNDERSIDE OF SPINNER

8. Put the bent paper clip through the hole in the four inch square, the paper washer and the pointer.

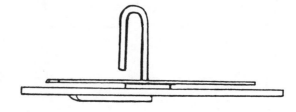

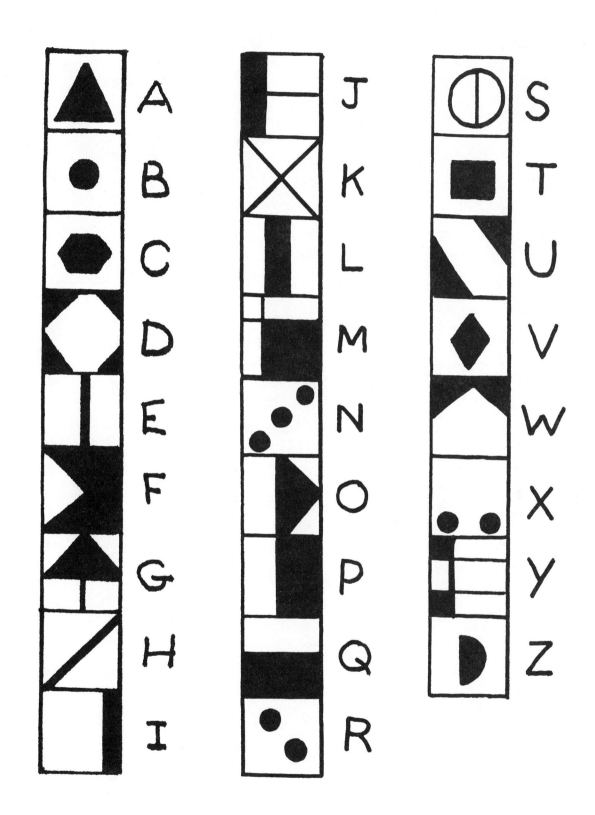

"GETTING THE ANGLE ON A MAP"

NORTH

DOWNTOWN
SAN FRANCISCO

SCALE: 3 INCHES = 1 MILE

FACES, NAMES, AND SHAPES

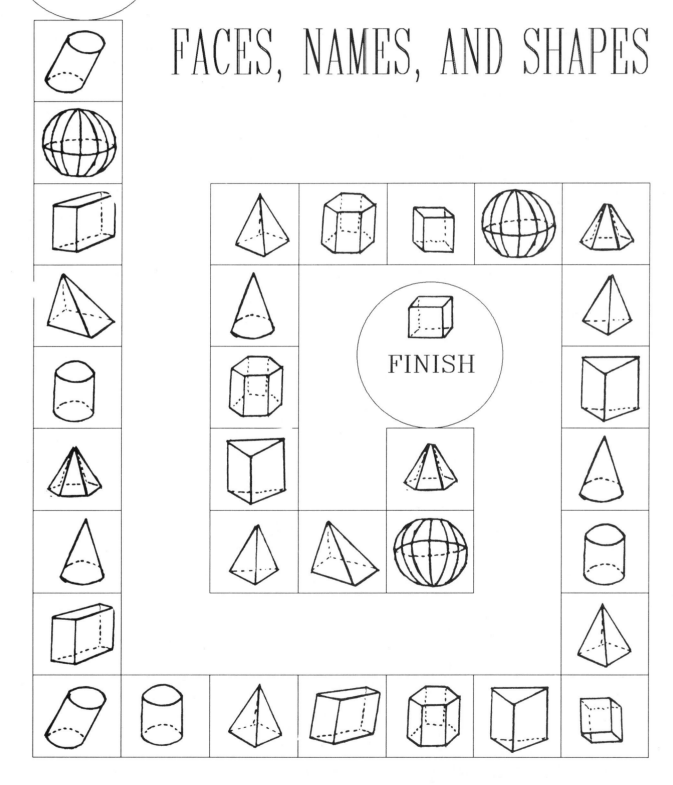

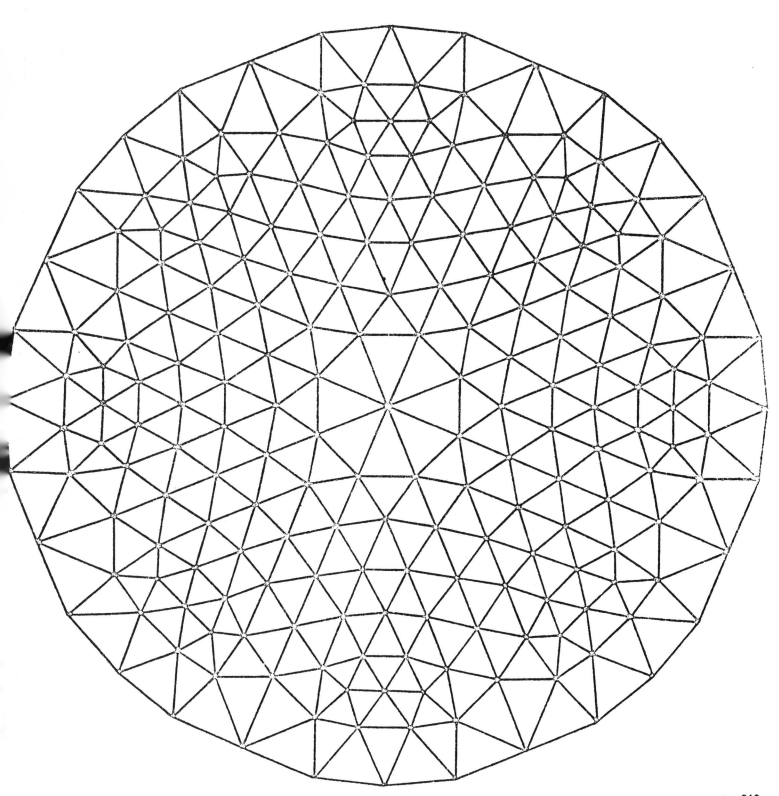